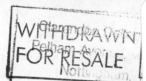

A Textbook of Anatomy and Physiology

Other titles published by Stanley Thornes (Publishers) Ltd:

THE PRINCIPLES AND PRACTICE OF PHYSICAL THERAPY
PRINCIPLES AND PRACTICE OF PERFUMERY AND COSMETICS
A TEXTBOOK OF HOLISTIC AROMATHERAPY

The Author: William Arnould-Taylor

Dr Arnould-Taylor has a distinguished record in the field of Physiology especially as applied to Physical Therapy. He is a member of "Epidemiology and Community Medicine" and founder member of "Clinical Forensic Medicine" at the Royal Society of Medicine.

He has devoted many years to research in areas relating to his special fields of interest with published scientific papers in the United Kingdom and America.

He is well known through his various textbooks, lectures and broadcasts in most English-speaking Countries throughout the world.

A Textbook
of
Anatomy and Physiology

by

William Arnould-Taylor

MSc PhD (Physiology)
Life Fellow, Royal Society of Medicine
Founder Member, Sections of Clinical Forensic Medicine
and Sports Medicine
Member, Section of Epidemiology and Public Health

Second Edition

Stanley Thornes (Publishers) Ltd

First published in 1977 by
Stanley Thornes (Publishers) Ltd
Ellenborough House
Wellington Street
CHELTENHAM GL50 1YW
United Kingdom

Second Edition 1988

96 97 98 99 00 / 10 9 8

British Library Cataloguing in Publication Data

Arnould-Taylor, W. E.
  A textbook of anatomy and physiology.
  2nd ed.
  1. Man. Anatomy
  I. Title
  611

ISBN 0–85950–937–0

Typeset by Quadraset Ltd, Midsomer Norton, Nr. Bath, Avon
Printed and bound at Redwood Books, Trowbridge, Wiltshire

Preface to Second Edition

Anatomy and Physiology are subjects of wide interest and it is hoped that this small textbook will prove useful to all those disciplines where a background knowledge of human structure and function is essential, as well as to those questing minds who wish to understand how their own body works.

I would like to place on record my appreciation of the help and guidance of my long-time collaborator Mrs Kim Aldridge M Phys.

William Arnould-Taylor

1988

Contents

Chapter 1

Introduction to Anatomy and Physiology

A review of standard textbooks on anatomy and physiology reveals that the majority are written with the medical student or nurse particularly in mind and few—if any—have been written specifically for the physical therapist. The result is that the physical therapist often has to wade through a good deal of material which is going to be largely valueless to him/her in the practice of the profession. The difference of approach between, say, a nurse on the one hand and a physical therapist on the other, is a basic one—the nurse is primarily concerned with what happens inside the body, that is beneath the skin, whereas the physical therapist is more concerned with the exterior of the body and those parts of the body which may be influenced from the exterior.

It has been found that very detailed textbooks on anatomy and physiology, though excellent for medical and nursing students, leave other students a little confused as to what is most suitable for them to select for the purposes of their studies. It is hoped, therefore, that this comparatively small section on anatomy and physiology will help to fill the gap. It makes no pretence of enveloping the whole of the subject matter of anatomy and physiology but aims to give the student a basic understanding of the relative parts of the body and their functions.

First of all the student should have a clear understanding of the relativity of the two terms—anatomy and physiology. Anatomy is normally defined as being the study of the structure of the body and the relationship of various parts one to another; whereas physiology is the study of the functions of those parts. For example—to say that the human heart is approximately 255g in weight, is somewhat pear-shaped in appearance and lies two-thirds to the left-hand side of the rib cage and one-third to the right, is to describe its anatomy; that is its weight, shape and position. On the other hand, this information tells us nothing at all about the function of the heart so we have to look at the physiology to learn that the heart is basically a pump which forces oxygenated blood around the body and, at the same time, circulates the venous or carbon dioxide loaded blood around the lungs for reoxygenisation. Therefore, by combining the knowledge of anatomy and physiology of the heart we are able to get a picture of what it looks like as well as what it does. In the ensuing chapters anatomy and physiology are dealt with together.

For the purpose of simplicity of learning—the body is divided into eight systems. In some textbooks these basic systems are sub-divided so that nine, ten or more systems are quoted. However, on closer investigation, it will be seen that these additional systems are really part of the basic systems. These we divide as follows:

(1) *The skeletal system* which, as its name implies, is the bony structure on which the other systems depend for support.
(2) *The muscular system.*
(3) *The vascular system* which includes the lymphatic system.
(4) *The neurological system* which covers all the nerves of the body as well as the brain.

These four systems we refer to as the *major systems* —not because they are more important than the systems which follow—but because they are the systems which envelop the whole of the body and are the ones which can most easily be affected by the professional skills of the physical therapist.

We now pass on to:

(5) *The digestive system.*
(6) *The respiratory system.*
(7) *The genito-urinary system* which includes the reproductive and kidney systems.
(8) *The endocrine system.*

These last four systems are referred to as the *minor systems*.

All professions have their own peculiar vocabularies —these are necessary for the accurate understanding of the subject matter—and the profession of medicine is no exception. It has a very wide vocabulary which requires a lifetime to master fully. There are, however, parts of this vocabulary which are essential to the physical therapist and these are dealt with in the form of a glossary at the end of each chapter so as not to interfere with the flow of the text of the lesson itself. The attention of students is, however, specially directed to the necessity of understanding this terminology because if the words are learnt in relation to the appropriate discussion it makes the subsequent learning that much easier.

HISTORICAL

It is not possible to trace the beginnings of the study of anatomy and physiology for these are lost in antiquity. The ancient Egyptians were famous for their embalming processes which must have involved a certain amount of knowledge of the anatomy of the human body and they had a system of medicine— traces of which survive until today. The R which a doctor writes at the top of his prescription is, in fact, the 'R' symbol for the Eye of Horus—the hawk- headed sun god who lost his eye in battle and had it restored by Thoth, the patron god of physicians. Thoth was one of the many gods invoked by doctors of ancient Egypt when administering their remedies.

About the same time, but in an entirely different part of the world, the Chinese were practising a form of medicine, acupuncture, which involved some 365 different needling points, but it was to the civilisation of the Greek period that we have to look for more detailed knowledge of human anatomy.

From this era we have the work of Hippocrates who is often referred to as the father of medicine and, in a rather different role, the name of Aristotle, who is generally acknowledged as being the founder of comparative anatomy.

During the Roman period which followed there was a medical school in Rome and the fine selection of surgical and dissecting instruments which have been preserved indicates a considerable knowledge of the structure of the human body.

It was in the 2nd century AD that Galen lived and his name is still remembered as being that of one of the greatest physicians and anatomists of antiquity. His work formed the basis of the European knowledge of anatomy for well over a thousand years, surviving through the Dark Ages into the Middle Ages when we see the beginning of the great Italian medical schools and universities such as Bologna and Padua. One of the 16th century graduates was Paracelsus von Hohenheim, a progressive medical teacher, who did much to alter the accepted ideas of his day. It was in 1543 that Vesalius published his first drawings of the structure of the human body and so paved the way for modern anatomy. Nearly a hundred years later, in 1628, Harvey announced his discovery of the role of the heart in the circulation of the blood through the lungs and the body, and in 1661 Malpighi discovered the capillary circulation and so completed the knowledge of how blood from arteries is returned to the heart by way of veins.

In the middle of the 18th century Auenbrugger of Austria invented percussion—a method by which doctors can diagnose the condition of the lungs. As a boy he had often watched his father tap barrels to see how much wine they contained and he applied the same technique to the chests of his patients. If they gave out hollow sounds similar to those of empty barrels he considered they were healthy whilst a muffled or high-pitched note indicated the presence of some unhealthy fluid.

At the end of the 18th century—1798—Jenner discovered that vaccination could be employed as a preventive of smallpox. Early in the 19th century the French physician, Rene Laennec, invented the stethoscope. He was attending a patient suffering from heart disease and, as she was rather obese, he decided that applying an ear direct to the chest (which was the usual method) would be of little use. He remembered that children sometimes amused themselves by playing with logs of wood, one child making tapping or scratching noises at one end and the other one listening at the other. So he rolled up a cylinder of paper and put one end of the stethoscope to the patient's chest and his ear to the other and found that he could hear the heart beating much more clearly than before. He then experimented with other materials until the stethoscope was invented.

The first real knowledge of the digestive system came in 1822 when a man by the name of Alexis St. Martin was wounded in the stomach in a brawl near Lake Michigan. He recovered but the wound left a permanent hole through which Dr. William Beaumont, U.S. Army, was able to watch how the stomach exuded the juices needed for digestion.

In the 1840s nitrous oxide or laughing gas was first used by a dentist in America for the extraction of teeth. This was quickly followed by the use of ether in hospital operating theatres which made possible a much more detailed study of anatomy. In 1867 Lister established the principles of antisepsis and in 1877 Pasteur demonstrated the role of germs in the causation of disease. In 1895 Röntgen discovered X-rays and in 1898 the Curies isolated radium. In 1904 Bayliss and Starling identified the first hormone. 1912 saw the discovery of vitamins by Frederick Gowland Hopkins whilst in 1928 Alexander Fleming discovered the antibiotic—penicillin—though this was not to come into medical use until about 1939.

Anatomy and physiology are subjects of continuous research and discovery and all the knowledge which we have accumulated in this century serves to indicate that we are only at the beginning of a complete understanding of these two subjects.

GLOSSARY

Glossary of general terms in Anatomy and Physiology.

The Anatomical Position — an erect position of the human body with arms by sides and the palms of the hands facing forward

Anterior — applies to the front of the body when in the erect position

Distal — the opposite of proximal and the part furthest away from the median line; so distal thigh will be at the knee end of the thigh

Dorsal — synonymous with posterior; normally used when describing the hand or the foot

Lateral — either side of the median line, e.g. the outer side of the arm will be its lateral aspect whilst the inner side is described as the medial aspect

Median Line — an imaginary line which runs through the centre of the body from the centre of the crown of the head ending up directly between the two feet

Morphology — the study of differences and resemblances in structure and form

Posterior — the back of the body when in the erect position

Proximal — a term of comparison applied to structures which are nearer the centre of the body or the median line, e.g. proximal thigh is the end of the thigh nearest to the centre of the body

Symmetrical — similar parts of the body, e.g. right and left ears, eyes, tibias, or limbs

Chapter 2

The Skeletal System

The skeleton provides the framework of the body and it has two principal functions. The first is that of *protection*, for example:

the skull protects the brain,

the rib cage protects principally the heart and the lungs,

the spinal column protects the spinal cord,

the pelvic bones provide a certain amount of protection for the viscera.

The second function is that of *locomotion* or *movement.*

The skeleton is made up of 206 bones though this figure varies slightly in different textbooks due to the fact that some authorities count the number of bones which are present in a young child whereas other authorities consider that only the bones of an adult should be counted, as by the time adulthood is reached certain childhood bones will have fused together.

Bone is a dry dense tissue composed of approximately 25% water, 30% organic material and 45% mineral. The mineral matter consists chiefly of calcium phosphate and a small amount of magnesium salts; these give the bone its rigidity and hardness. The organic matter consists of fibrous material which give the bone its toughness and resilience. There are five classifications of bone:

(1) *Long bones* e.g. the femur or thigh bone, the longest and strongest bone of the body.

(2) *Short bones* e.g. metatarsal bones.

(3) *Flat bones* e.g. frontal bone of the head.

(4) *Irregular bones* e.g. the vertebrae.

(5) *Sesamoid bones*—rounded masses found in certain tendons of muscles, the best example being the patella or knee cap.

A long bone normally consists of marrow surrounded by a spongy bone layer which, in turn, is surrounded by a compact bone layer and finally by a hard outside covering known as the periosteum.

In addition to the two principal functions of the skeleton, individual bones serve other purposes such as the attachment of tendons and muscles, and the formation of red blood cells and some white blood cells in the bone marrow.

A *joint* is formed where two bones meet. Joints may be divided according to their mobility, into three types:

(1) *Fixed joints or synarthroses* provide no movement, for example the sutures between the

THE SKELETON

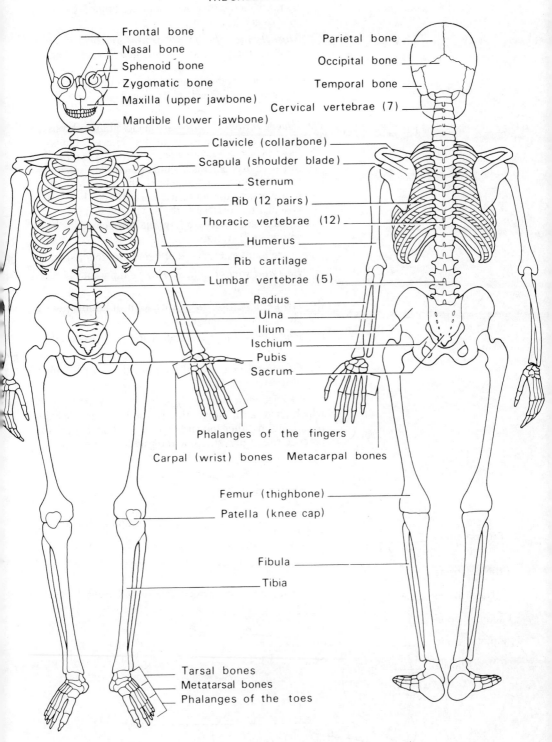

Frontal bone
Nasal bone
Sphenoid bone
Zygomatic bone
Maxilla (upper jawbone)
Mandible (lower jawbone)

Parietal bone
Occipital bone
Temporal bone
Cervical vertebrae (7)

Clavicle (collarbone)
Scapula (shoulder blade)
Sternum
Rib (12 pairs)
Thoracic vertebrae (12)
Humerus
Rib cartilage
Lumbar vertebrae (5)
Radius
Ulna
Ilium
Ischium
Pubis
Sacrum

Phalanges of the fingers

Carpal (wrist) bones   Metacarpal bones

Femur (thighbone)
Patella (knee cap)

Fibula
Tibia

Tarsal bones
Metatarsal bones
Phalanges of the toes

skull bones. There is fibrous tissue between the bones, which either overlap or are fitted together in a jagged line.

(2) ***Slightly movable joints or amphiathroses*** are found in the pelvis (*symphysis pubis*), sacro-iliac joint and the joints at both ends of the clavicle. The bones are held together by strong ligaments and separated by pads of fibrocartilage (*cartilaginous joints*).

(3) ***Freely movable joints*** are enclosed in a fibrous capsule, supported by ligaments. This capsule is lined by a *synovial membrane* with *synovial fluid* in the cavity. This is a whitish fluid, not unlike raw egg-white in consistency, which acts like oil in a machine to reduce friction between the articulating surfaces of the joint. The bone surfaces are covered by *hyaline* cartilage for smoother operation.

There are four main groups of freely movable joints:

(a) ***Ball and socket articulations***—hip joint, shoulder joint

(b) ***Hinge articulations***—knee joint (full hinge), elbow joint (partial hinge)

(c) ***Pivot articulations***—radius and ulna joints, axis joint of head

(d) ***Gliding joints***—tarsal joint of ankle, carpal joint of wrist.

The knee joint is the only articulation in the body which forms a full hinge, that is, the bones are capable of moving in either forward or backward directions. In practice this is prevented by the patella or kneecap

THE SKULL

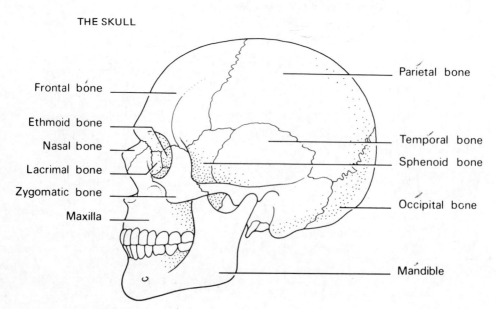

Frontal bone

Ethmoid bone

Nasal bone

Lacrimal bone

Zygomatic bone

Maxilla

Parietal bone

Temporal bone

Sphenoid bone

Occipital bone

Mandible

which fits into the hinge rather like a doorstop or wedge. When the patella is not present, for example when it has been broken in an accident, the lower leg comes forward.

The capsule of the freely movable joints possesses small sacs containing a clear, viscous fluid. These structures which are called *mucous bursae* pad the joint or articulation rather like water cushions. If the synovial membrane becomes inflamed this is known as *synovitis,* a well known example being tennis elbow. If the bursae become inflamed this is known as *bursitis;* the best known example of this is housemaid's knee—an occupational hazard for people whose work involves a good deal of kneeling.

Distribution of Bones in the Skeleton

22 bones form the skull bones, 8 of the cranium:

1 frontal bone forming the forehead
2 parietal bones forming the top and sides of the cranium
1 occipital bone
2 temporal bones
1 sphenoid bone
1 ethmoid bone

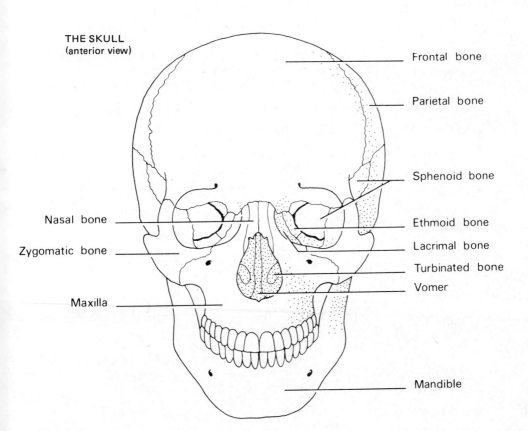

THE SKULL
(anterior view)

Frontal bone
Parietal bone
Sphenoid bone
Nasal bone
Ethmoid bone
Zygomatic bone
Lacrimal bone
Turbinated bone
Vomer
Maxilla
Mandible

and 14 bones of the face—of which the principal bones are:

> the superior maxilla or upper jaw
> the mandible or lower jaw (the jawbone which moves)
> 2 zygomatic or cheek bones
> 2 nasal bones which form the bridge of the nose
> 2 lacrimal bones.

25 bones form the thorax (or chest):

> the sternum or breast bone
> 12 pairs of ribs.

The first 7 pairs are known as *true ribs* because each rib is joined to the sternum directly. The next 5 pairs (8–12th) are known as *false ribs* because they do not join the sternum directly. The 8th, 9th and 10th ribs fuse with the rib immediately above, while the 11th and 12th pairs (*floating ribs*) only partly surround the circumference of the thorax and are unattached in front.

33 bones form the spine:

> 24 *true* or *movable* vertebrae, separated by pads of fibrocartilage
> 9 *false* or *fixed* vertebrae, closely fused together with no movement between them.

From the top of the spine downwards there are:

> 7 cervical vertebrae—the first the atlas bone, second the axis bone
> 12 thoracic vertebrae
> 5 lumbar vertebrae
> 5 sacral vertebrae—fused to form the sacrum
> 4 coccygeal vertebrae—fused to form the coccyx.

4 bones form the shoulder girdle:

> 2 clavicles or collar bones
> 2 scapulae or shoulder blades.

60 bones form the upper limbs, 30 bones in each whole arm:

> 1 humerus or upper arm
> 1 radius—the outer bone of the forearm
> 1 ulna—the inner bone of the forearm
> 8 carpal bones forming the wrist
> 5 metacarpal bones forming the hand, followed by 14 phalanges or finger bones.

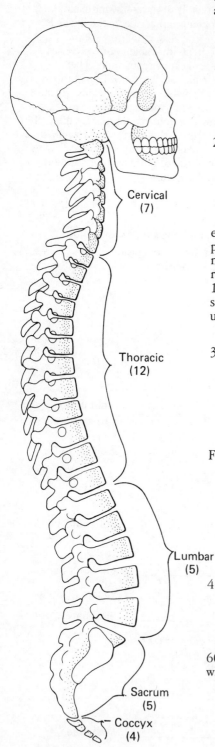

Cervical (7)

Thoracic (12)

Lumbar (5)

Sacrum (5)

Coccyx (4)

BONES OF THE SPINE

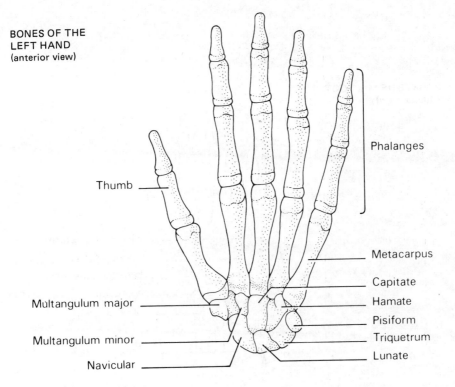

**BONES OF THE
LEFT HAND**
(anterior view)

Phalanges

Thumb

Metacarpus

Capitate

Hamate

Multangulum major

Pisiform

Multangulum minor

Triquetrum

Navicular

Lunate

RIGHT FEMUR
(anterior view)

Head

Neck

Front of shaft

Patella surface

The pelvis is formed by 4 bones:

the right and left innominate bones (hip bones)
the sacrum and coccyx already referred to as part
of the spinal vertebrae.

Each innominate bone consists of:

the ilium—upper portion
the ischium—rear portion
the pubis—front portion.

60 bones go to form the lower limbs, 30 in each
whole leg:

the femur or thigh bone
the patella or kneecap
the tibia or shin bone
the fibula or brooch bone
7 tarsal bones of the ankle
5 metatarsal bones of the foot
14 phalanges of the toes.

BONES OF THE FOOT
(dorsal view)

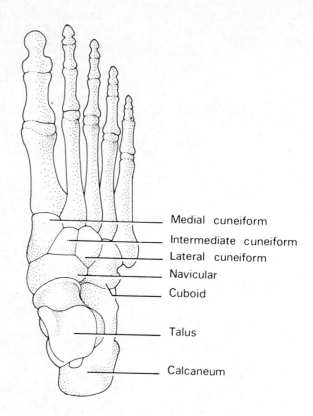

Medial cuneiform

Intermediate cuneiform

Lateral cuneiform

Navicular

Cuboid

Talus

Calcaneum

In addition to the bones enumerated there is one hyoid bone which lies in the front upper part of the neck and is detached from the skeleton.

Spinal Curvature

Reference to the illustration of the spine shows that it has two natural curves—the slightly outward curving upper part of the spine being in the thoracic region and the inward curving of the spine being in the lumbar region. These natural curvatures can be exaggerated by three basic causes:

(1) *Congenital*—present at the time of birth or arising as a direct result of hereditary factors.

(2) *Traumatic*—resulting from accidents.

(3) *Environmental*—resulting from bad posture and often closely allied to the type of work in which the subject is engaged.

There are three types of curvature:

(1) Exaggerated outward curvature of the spine in the thoracic region is referred to as *kyphosis*.

(2) An inward exaggeration of the spine in the lumbar region is called *lordosis*.

(3) A lateral curvature of the spine is known as *scoliosis*. Scoliosis may occur at any part of the spine and is quite often associated with one of the other curvatures, e.g. the 'Hunchback of Notre Dame' suffered from kyphosis and scoliosis of the thoracic region.

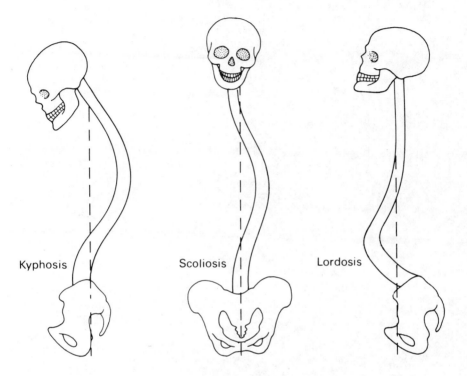

Kyphosis   Scoliosis   Lordosis

CURVATURE OF THE SPINE

Fractures

When a bone breaks it is referred to as a fracture; fractures are divided into a number of categories:

(1) *A simple fracture* is when a bone breaks in one place and no serious damage is done to the surrounding tissues.

(2) *A complicated fracture* is when the bone is broken and the break causes injury of the surrounding soft tissue.

(3) *A compound fracture* is when the bone is broken and one or both ends protrude through the external surface of the body, that is, through the skin.

(4) *A comminuted fracture* is where the bone is broken in a number of places.

(5) *An impacted fracture* is where the bone is broken and one end is driven into the other.

(6) *A greenstick fracture* is an incomplete fracture of a long bone as seen in young children.

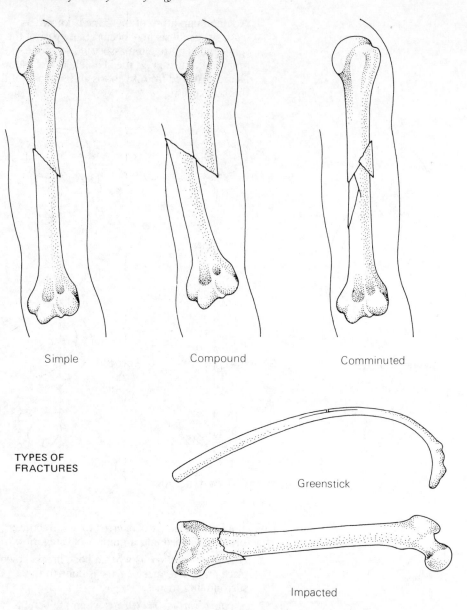

Simple   Compound   Comminuted

TYPES OF FRACTURES

Greenstick

Impacted

DISEASES OF THE SKELETON

The skeleton is subject to many diseases, most of which do not come within the purview of this text-book, but mention must be made of arthritis because it is such a universal disease. Technically arthritis is an inflammation of a joint but this is generally interpreted as being a rheumatic affection of the joint, i.e. the inflammation is caused by the type of rheumatism which primarily affects the skeletal system as distinct from fibrositis and neuritis which are dealt with in later chapters.

There are many types of arthritis—among the more common being:

Mono-articular arthritis, as the name implies, is a type of arthritis which involves only one joint.

Poly-articular arthritis attacks a number of joints, usually associated joints, i.e. both hips or both knees or, in some cases, all the joints of one leg or arm.

Osteo- or degenerative arthritis is a chronic joint disease characterised by loss of some of the joint's cartilage and some spur formation.

Rheumatoid arthritis is a chronic arthritis usually associated with hormone deficiency.

GLOSSARY

Appendicular Skeleton	skeleton of the upper and lower limbs and their girdles
Axial Skeleton	skeleton of the head and trunk
Cancellous Tissue	characterised by a latticed structure as seen in the spongy tissue of bones
Cartilage	a substance similar to bone but not as hard. It acts as a cushion between bones and also gives shape to nose and ears.
Dislocation	occurs when force is applied to a joint and is greater than that necessary to produce a strain. It particularly applies to the ball and socket joints as the ball is forced out of the socket. When dislocated bones are returned to their proper position this is referred to as *reduction.*
Gouty Arthritis	occurs in any part of the body but is popularly associated with the big toe. It results from urate crystals (chalky salts of uric acid) being deposited in and around the cartilage. This form of arthritis is much more common in men than women.
Orthopaedics	branch of surgery concerned with corrective treatment of skeletal system
Osteo	referring to bone
Periosteum	the hard membrane adhering to a bone and forming a protective cover. It contains blood vessels supplying blood to the bone and at its deepest layers are the bone-forming cells—osteoblasts.
Rickets	a calcium deficiency disease of children usually evidenced by misshapen bones
Spondylitis	a type of arthritis which attacks the spinal vertebrae; the severest form is *ankylosing spondylitis* where bone and cartilage fuse resulting in complete immobility

Chapter 3

The Muscular System

The main framework of the skeleton of the body is covered by muscles. These are responsible for 50% of our body weight and their function is to permit movement, for which purpose they are, in most cases, attached to bones.

There are two types of muscle—*voluntary* and *involuntary.* The voluntary muscles, such as those used in walking or writing, are muscles which are under conscious control. Involuntary muscles are those which are involved in movements of the heart, respiration, digestion and so on, and are outside conscious control.

A section of *voluntary* muscle shows it to be of striped and striated (cross-banded) tissue whilst *involuntary* muscles have slender, smooth types of cells without cross stripes and are therefore usually referred to as smooth muscles.

A section of *cardiac* muscle tissue shows that whilst it is involuntary muscle it has characteristics which bear a superficial resemblance to voluntary muscle tissue though the fibres are smaller than those of voluntary muscles and the striae are not so well marked.

A muscle consists of a number of contractile or elastic fibres bound together in bundles. The bundles are, in turn, bound together by a thick band usually spindle-shaped and always contained in a sheath. This sheath is extended at the end to form strong fibrous bands known as the tendons by means of which the muscles are fastened to the bones.

CROSS-SECTION
THROUGH MUSCLE

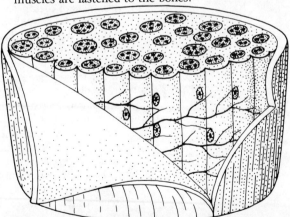

The muscles which are responsible for skeletal movements have two points of attachment—the point of origin is the bone to which they are attached and which they *do not* move and the point of insertion is the bone to which they are attached and which they *do* move, e.g. the biceps of the arm has its points of attachment—that is points of origin—at the shoulder end of the arm, whilst the points of insertion are the radius of the lower arm—therefore it is the lower arm that is activated by the biceps.

A muscle receives its stimulus from a motor nerve and in response to this stimulus it shortens its length so that, *in action,* a muscle always contracts. Muscles of the body normally work in pairs, the one that is the prime mover at any given time being known as a *synergist* and the one which holds it in check as the *antagonist.* This means that the muscles of the body—i.e. the voluntary muscles—are never completely at rest. They are normally in a condition of slight tension or contraction and this we call muscular tone.

For example, when we bend our forearm the muscles on the front of the arm contract whilst, at the same time, the muscles on the back of the arm relax gradually to maintain balance. In this instance the muscles at the front of the arm are the synergists whilst the ones at the back are the antagonists. However, to reverse this position and straighten the arm out again the muscles on the back of the arm become the synergists and those on the front the antagonists.

Altogether there are some 640 named muscles in the body but there are many, many thousands of unnamed ones—each hair on the surface of the body having a tiny muscle attached to it. When a person gets chilled or frightened and has what are known as 'goose pimples'—the little lumps on the skin are due to the tiny muscles of the skin pulling the hair erect. Muscles are well supplied with arteries to bring them food for fuel and repair, and oxygen for combustion of the fuel, and with veins which carry away the waste products of their activities, such as carbon dioxide.

Muscular activity contributes materially to the internal heat of the body and when there is a danger of this reaching too low a level a person shivers. This is an involuntary action making the muscles work in order to generate more heat. Muscles are, in turn, responsive to exterior heat so that exposure of the skin to cold air increases muscle tone whereas considerable heat, e.g. a hot bath, has a relaxing effect on muscles. About 30% of the energy produced in muscle activity results in work and the remaining 70% is released as heat which warms the body, particularly the blood.

As previously mentioned in this chapter—muscle contraction occurs as the result of a stimulus which it receives from a motor nerve. This nerve stimulus sets up chemical changes in the muscles. These changes include the breaking down of glucose, glycogen and fat, which, in turn, liberate the energy required for contraction. In the process of contraction there are some waste products which are excreted from the muscles by the venous system. However, if at any one time the muscular activity is so great as to produce more waste products than the venous and lymph systems are able to cope with, then some waste products remain in the muscle or between the muscle fibres and give a feeling of stiffness—that is the fibres are no longer easily able to slide one over the other.

Muscles are put into groups according to the functions which they perform:

An extensor extends a limb.

A flexor flexes a limb.

An adductor bends a limb towards the median line.

An abductor takes a limb away from the median line.

A sphincter surrounds and closes an orifice or opening.

A supinator turns a limb to face upwards.

A pronator turns a limb to face downwards.

Rotators rotate a limb.

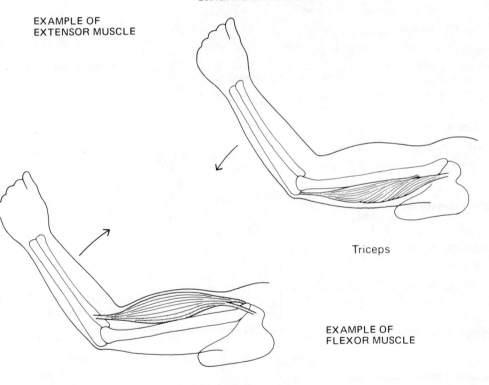

EXAMPLE OF
EXTENSOR MUSCLE

Triceps

EXAMPLE OF
FLEXOR MUSCLE

Biceps

The following is a short list of some of the principal muscles of the body. This is by no means a complete list and students who wish to study the subject in greater depth as well as to learn the origins and insertions of muscles are referred to one of the standard anatomical textbooks dealing with this subject. But this list should cover most, if not all, of the muscles that the physical therapist is likely to have to deal with.

Head and Neck

Name	Action
Frontalis (or epicraneas)	elevates eyebrows and draws scalp forward
Orbicularis oculi	closes eyelids
Orbicularis oris	puckers mouth
Masseter	muscle of mastication, closes mouth, clenches teeth

MUSCLES OF THE BODY
(posterior view)

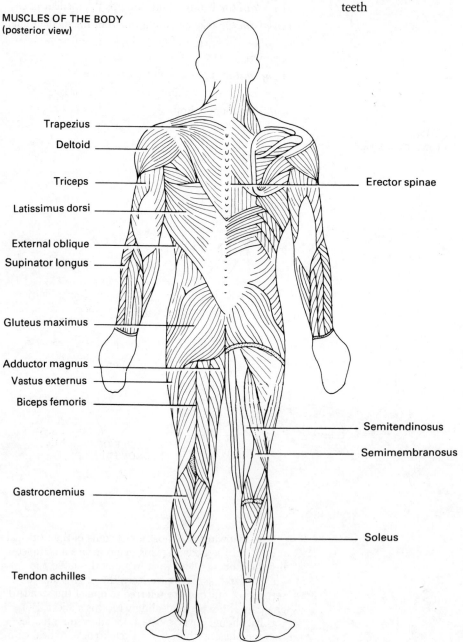

Trapezius

Deltoid

Triceps

Latissimus dorsi

External oblique

Supinator longus

Gluteus maximus

Adductor magnus

Vastus externus

Biceps femoris

Gastrocnemius

Tendon achilles

Erector spinae

Semitendinosus

Semimembranosus

Soleus

Name	Action
Buccinator	compresses cheeks and retracts angle of mouth
Sterno-cleido mastoid (Sterno-mastoid)	flexes head and turns from side to side
Platysma	muscle of facial expression

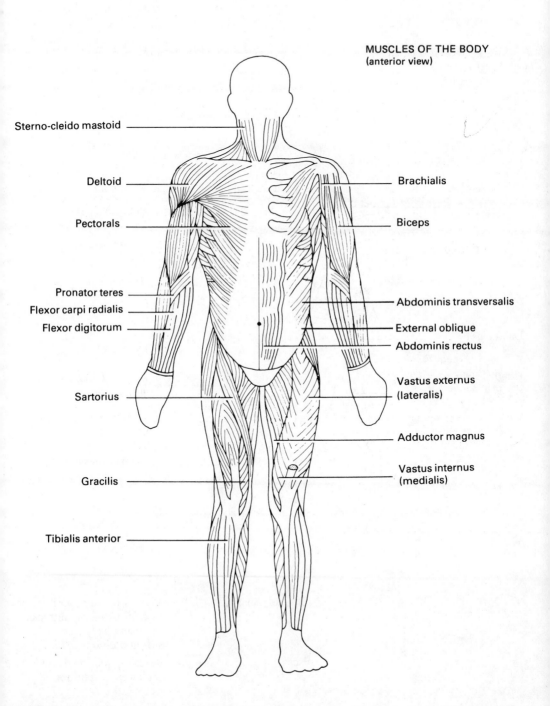

MUSCLES OF THE BODY
(anterior view)

Sterno-cleido mastoid

Deltoid

Pectorals

Pronator teres

Flexor carpi radialis

Flexor digitorum

Sartorius

Gracilis

Tibialis anterior

Brachialis

Biceps

Abdominis transversalis

External oblique

Abdominis rectus

Vastus externus
(lateralis)

Adductor magnus

Vastus internus
(medialis)

Trunk of Body

Name	Action
Trapezius	rotates inferior angle of scapula laterally, raises shoulder, draws scapula backwards
Erector spinae	extends vertebral column
Latissimus dorsi	adducts the shoulder and draws the arm backwards and downwards
Serratus magnus	draws the scapula forward

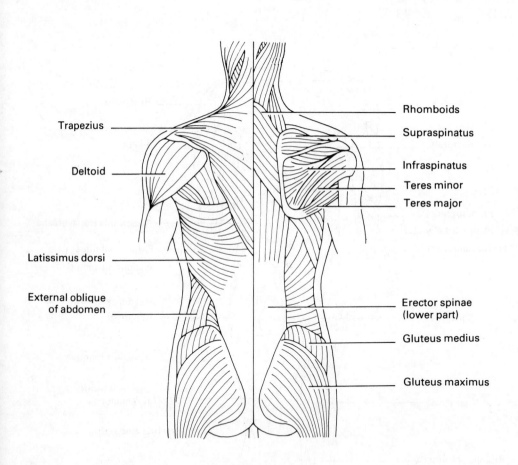

Trapezius

Deltoid

Latissimus dorsi

External oblique
of abdomen

Rhomboids

Supraspinatus

Infraspinatus

Teres minor

Teres major

Erector spinae
(lower part)

Gluteus medius

Gluteus maximus

MUSCLES OF THE BACK

Gluteus maximus	extends hip joint and extends trunk on buttocks in raising body from sitting position
Psoas	flexes hip joint and trunk on lower extremities

Name	Action
Pectoralis major	Flexes shoulder joint, depresses shoulder girdle, adducts and rotates humerus
Abdominis obliquus (internal and external oblique)	supports abdominal viscera and flexes vertebral column
Abdominis transversalis (transversus abdominis)	supports abdominal viscera and flexes vertebral column
Abdominis rectus	supports abdominal viscera and flexes vertebral column

The Arms

Deltoid	abduction of the humerus to right angle
Biceps brachialis	flexes and supinates forearm
Triceps brachialis	extends elbow joint
Brachialis anticus	flexes elbow joint
Coraco brachialis	flexes and adducts humerus
Brachio radialis (supinator longus)	flexes elbow joint
Pronator teres (pronator radii terres)	pronates forearm
Supinator (Supinator radii breves)	supinates forearm
Flexor carpi radialis	flexes wrist joint
Extensor carpi (radialis longus)	extends wrist
Flexor carpi ulnaris	flexes wrist joint
Extensor carpi ulnaris	extends wrist joint
Flexor sublimis digitorum	flexes fingers
Extensor sublimis digitorum	extends fingers

The Legs

Rectus femoris (Quadriceps)	extends knee joint
Vastus lateralis (externus) (Quadriceps)	extends knee joint
Vastus medialis (internus) (Quadriceps)	extends knee joint
Vastus intermedius (Quadriceps)	extends knee joint
Sartorius	flexes hip and knee joints and rotates femur
Adductor magnus, longus, and brevis	adduct thigh
Biceps femoris (ham string)	flexes knee joint

Name	*Action*
Semitendinosus	flexes knee joint and extends hip joint
Semimembranosus	flexes knee joint and extends hip joint
Gracilis	adducts femur and flexes knee joint
Gastrocnemius	flexes ankle and knee joint
Tibialis anticus	flexes and inverts foot
Peroneus longus	inverts and flexes foot and supports arches
Flexor digitorum longus	flexes toes
Extensor digitorum longus	extends toes
Tendon of achilles	assists in flexion of the foot

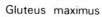

Gluteus maximus

Biceps femoris

Semitendinosus

Gracilis

Semimembranosus

Sartorius

Gastrocnemius

Soleus

MUSCLES OF THE LEG
(posterior view)

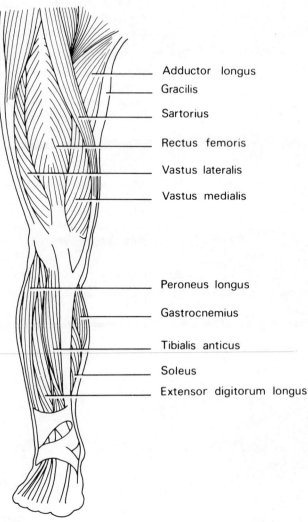

Adductor longus

Gracilis

Sartorius

Rectus femoris

Vastus lateralis

Vastus medialis

MUSCLES OF THE LEG
(anterior view)

Peroneus longus

Gastrocnemius

Tibialis anticus

Soleus

Extensor digitorum longus

COMMON DISEASES OR CONDITIONS AFFECTING THE MUSCULAR SYSTEM

One of the most common diseases of the system is *fibrositis* which technically means inflammation of soft tissue and is a term which is now generally applied to a rheumatic affection of the muscles—a condition in which there is a build-up of urea and lactic acid inside the muscle to the extent of causing stiffness and pain.

A well known example of this disease is *lumbago* or fibrositis of muscles in the lumbar region. *Torticollis* or 'wry neck' is another condition which has much in common with muscular fibrositis. In this case the muscle concerned is the sterno-cleido mastoid muscle of the neck which, in a state of contraction, causes the head to take up an abnormal position.

Another common condition is that of *cramp*. This is a localised painful contraction of one or more muscles, which has a number of causes, the most usual being that of vigorous exercise; but it also occurs in certain metabolic disorders, e.g. when there is a sodium depletion or water depletion. It is for the purpose of avoiding cramp that copious quantities of salted water are given to people who work in intense heat—for example, people who look after furnaces at steelworks.

There are also disease conditions which affect muscles though their cause is to be found in one of the other systems. For example, *poliomyelitis*—commonly called 'polio'—which arises in the neurological system and multiple or disseminated *sclerosis* which also arises in the neurological system, though both these disease conditions profoundly affect the body's musculature.

GLOSSARY

Atony	abnormally low degree of tonus or absence of it
Atrophy	reduction in the size of a muscle which previously reached a matured size; popularly referred to as wastage
Cramp	painful involuntary contraction of muscle
Fascia	the sheath or membrane covering a muscle
Ganglion	a cystic swelling which occurs in association with a joint or tendon sheath. Ganglia most commonly occur on the back of the wrist
Myology	the science of muscles
Myositis	inflammation of a muscle
Ligaments	bands of fibrous tissue which help to bind the bones of joints together
Rupture	a tearing or bursting of the fascia or sheath which surrounds the fibres of the muscle
Spasm	a sudden muscular contraction
Sprain	an injury to a ligament
Strain	an injury to a muscle or its tendon
Tendon	a band of fibrous tissue forming the end of a muscle and attaching it to the bone
Tonus	muscle tone
Viscera	the contents of the abdominal cavity

Chapter 4

The Vascular System

This system which is sometimes called the circulatory system consists of the heart, blood vessels, blood, lymphatic vessels and lymph.

The centre of this system is the heart, which is a muscular organ that rhythmically contracts, forcing the blood through a system of vessels. The heart weighs approximately 255 g in a fully grown adult and lies one-third to the right and two-thirds to the left of the thoracic cavity. At birth it beats about 130 times a minute, at six years about 100 times a minute, reducing in adult life to between 65 and 80 beats a minute with an average somewhere around 70. During the 24 hour period an adult human heart pumps 36 000 litres of blood through the 20 000 km of blood vessels.

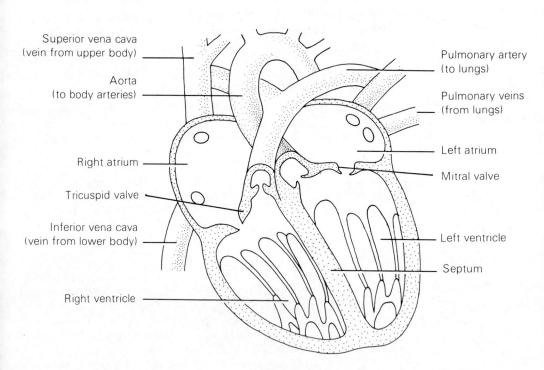

Superior vena cava (vein from upper body)

Aorta (to body arteries)

Right atrium

Tricuspid valve

Inferior vena cava (vein from lower body)

Right ventricle

Pulmonary artery (to lungs)

Pulmonary veins (from lungs)

Left atrium

Mitral valve

Left ventricle

Septum

THE HEART

The heart is divided into four chambers. These are the right and left *atria (or auricles)*, in the upper part of the heart, and the right and left *ventricles* in the lower part. The right side of the heart is divided from the left by a solid wall or *septum* which prevents the venous blood in the right side coming into contact with the arterial blood on the left side of the heart.

Circulation is divided into two principal systems: the *general* or *systemic* circulation (around the body); and the *pulmonary* circulation (to and from the lungs).

The general circulation includes two special branches: the *portal* circulation, which conveys blood from the digestive organs to the liver; and the *coronary* circulation, which supplies the heart muscle.

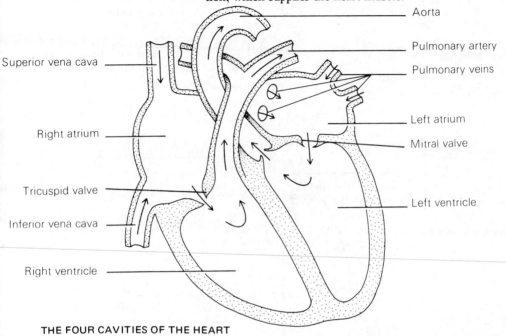

Aorta
Pulmonary artery
Pulmonary veins
Superior vena cava
Left atrium
Right atrium
Mitral valve
Tricuspid valve
Inferior vena cava
Left ventricle
Right ventricle

THE FOUR CAVITIES OF THE HEART

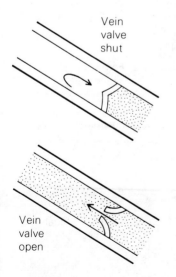

Vein valve shut

Vein valve open

VEINS
(convey blood to the heart)

Blood vessels which proceed from the heart are known as *arteries*. They generally carry oxygenated blood (the exception being the pulmonary artery). They are large, hollow, elastic tubes which gradually decrease in diameter as they spread through the body. These smaller vessels or *arteries* finally become very fine hairlike vessels known as *capillaries*.

Blood vessels which proceed towards the heart are known as veins. They generally carry deoxygenated blood (the exception being the pulmonary vein). They are elastic tubes with valves which prevent a backward flow of blood.

The veins empty the deoxygenated blood into the right atrium of the heart through the *inferior* and *superior vena cava*. The blood flows through the *tricuspid* valve to the right ventricle and is pumped to the lungs via the pulmonary artery. This is the only artery in the body to carry deoxygenated blood.

The blood is reoxygenated in the lungs and returns to the left atrium of the heart through the pulmonary veins. These are the only veins to transport oxygenated blood. The blood flows into the left ventricle through the *mitral* valve and is pumped to the body through the *aorta*.

The aorta is the largest artery in the body. It divides into two branches: the ascending aorta, supplying the arms and head; and the descending aorta, supplying the lower part of the body.

The descending aorta passes from the thorax through the diaphragm to the abdomen, where it is called the *abdominal aorta*. The *coeliac axis*, which branches off the abdominal aorta, supplies the stomach, liver and spleen. Below this the *renal* arteries branch off to the kidneys and the *mesenteric* arteries to the intestines. Finally the abdominal aorta branches into two *iliac* arteries which run into the pelvis. The

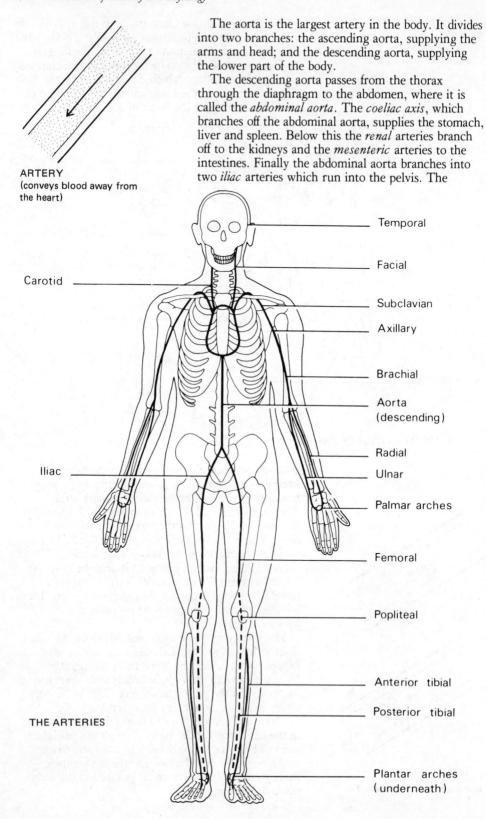

ARTERY
(conveys blood away from the heart)

Temporal

Facial

Carotid

Subclavian

Axillary

Brachial

Aorta
(descending)

Radial

Iliac

Ulnar

Palmar arches

Femoral

Popliteal

Anterior tibial

Posterior tibial

THE ARTERIES

Plantar arches
(underneath)

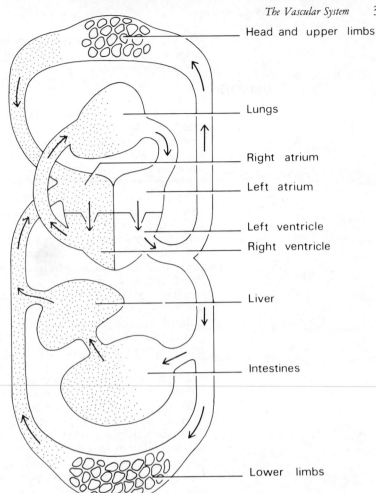

Head and upper limbs

Lungs

Right atrium

Left atrium

Left ventricle

Right ventricle

Liver

Intestines

Lower limbs

DIRECTION OF
CIRULATION

internal iliac artery supplies the reproductive organs while the *external iliac* artery becomes the *femoral* artery which is the main artery of the lower limb.

The femoral artery supplies the thigh muscles and becomes the *popliteal* artery at the knee. This divides into the *anterior* and *posterior tibial* arteries. The anterior tibial artery supplies the front of the leg and is continued to the foot as the *dorsalis pedis* artery. The posterior tibial artery supplies the back of the leg and reaches the sole of the foot as the *plantar artery* which forms the *plantar arch*.

Two *coronary* arteries branch off the ascending aorta, which then passes upwards as the *innominate* artery. This divides into the *subclavian* and *carotid* arteries.

The subclavian artery passes behind the clavicle and enters the armpit where it becomes the *axillary* artery. The *brachial* artery continues for the length of the upper arm until the elbow where it divides into the *radial* and *ulnar* arteries, culminating in the *palmar arches* in the hand.

The carotid artery passes upwards to the neck and has four main branches, the *facial, temporal, occipital* and *maxillary* arteries.

BLOOD

Blood is alkaline in reaction and amounts to approximately 5–6 litres in the average adult. It is complex in nature but has four principal constituent parts— *plasma, erythrocytes* or red corpuscles, *leucocytes* or white corpuscles and *platelets*.

Plasma provides the liquid basis of the blood. This is a clear, straw-coloured liquid which holds various substances in solution. These include sugar, urea, amino acids, mineral salts, enzymes, etc.

Erythrocytes or red corpuscles (corpuscles is Latin for little bodies) are inert biconcave discs, they get their colour from haemoglobin which has the ability to absorb oxygen (when it becomes *oxy-haemoglobin* which is bright red in colour) and carbon dioxide (when it becomes *carboxy-haemoglobin* which becomes very dark red, bordering on a muddy brown colour). The average life span of an erythrocyte is 120 days; they are produced mainly in red bone marrow and their eventual disintegration takes place in the spleen, and is finally completed in the liver.

In health, the erythrocytes total about 5 million per cubic millimetre of blood which gives a total of somewhere in the region of 25 billion in a human adult. If these cells were placed end to end they would form a ribbon sufficiently long to encircle the world more than four times. These cells are the body's transporters; they carry oxygen to all parts of the body and on their return journey pick up waste products, primarily carbon dioxide.

Leucocytes or phagocytes (white corpuscles) are larger than erythrocytes and have an irregular shape and a nucleus. They are produced in the bone marrow and, in health, they total about 8000 per cubic millimetre. They are the protectors or soldiers of the body; their chief role is to protect the body against infection by their power of ingesting bacteria—a process which is known as *phagocytosis*. When the body is subject to serious infection the leucocytes increase rapidly by a process of division known as mitosis.

Then we have platelets or thrombocytes. These average 250 000 per cubic millimetre of blood. They are derived from large multinucleated cells in the bone marrow and are essential to the blood for coagulation, i.e. clotting.

Blood Types

The existence of human blood types was established by Karl Landsteiner in 1902 when he began a study to determine why fatalities occurred following some blood transfusions. He discovered that the cause was incompatibility between the blood of the donor and the blood of the recipient.

Arising from this work came the Landsteiner Class-ification of Blood Groups which classified blood into the four types A, B, AB and O.

Type O is called the universal donor because it may give blood to all blood types but it can only receive from type O. On the other hand, type AB is called the universal recipient because it can receive from any group but can only give to the AB group. Type A can give to both A and AB and receive only from types A and O. Type B can give to type B and AB and receive only from types B or O.

In 1940 Landsteiner and A S Weiner recognised the Rh factor, a substance found in red blood cells. This was discovered during their experiments with rhesus

LYMPH NODES

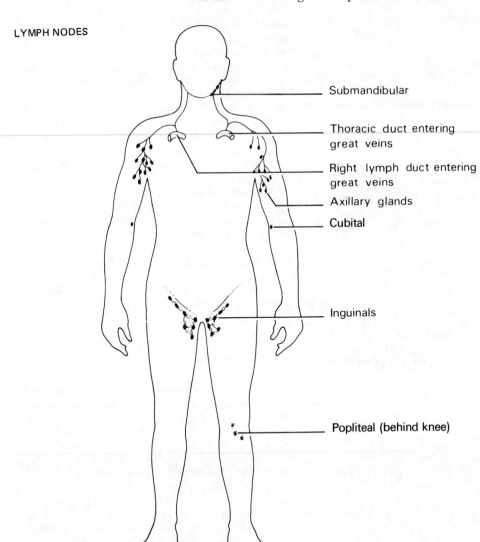

Submandibular

Thoracic duct entering great veins

Right lymph duct entering great veins

Axillary glands

Cubital

Inguinals

Popliteal (behind knee)

monkeys, hence the name rhesus or the abbreviation Rh. It is estimated that 85% of white people have Rh positive factor and the other 15% are Rh negative.

THE LYMPHATIC SYSTEM

This is a secondary circulation intertwined with the blood circulation. The basic material of the lymphatic system is the *lymph* which is plasma after it has been exuded from the capillaries. It gives nourishment to the tissue cells and in return takes away their waste products. The liquid is drained off by tiny lymphatic vessels which join together to form larger lymph vessels and, as these lymph vessels convey lymph towards the heart, they are supplied with valves in much the same way as veins. Along their course towards the heart there are receiving or reservoir areas known as *lymph nodes*. They vary in size from pin head to a small almond. The purpose of these lymph nodes is to filter the lymph as it passes through and, in this way, to help prevent infection passing into the blood stream and to add *lymphocytes* to the lymph.

Eventually all lymph passes into two principal lymph vessels, the *thoracic duct* and the right *lymphatic duct*, which open into the blood stream at the junctions of the right and left internal, jugular and subclavian veins where it becomes part of the general systemic circulation again.

There are approximately 100 of these lymphatic nodes scattered throughout the body along the line of the lymphatic vessels. The most common superficial ones are the *inguinals* in the groin, the nodes in the *popliteal fossa* or depression behind the knee, the *supratrochlea* in the crutch of the elbow, the *axillary glands* in the armpit, the *supraclavicular glands*, the *submandibular glands* underneath the mandible and the *cervical* and *occipital glands*. These superficial glands are the ones which swell when an infection is present in that part of the body.

CONDITIONS, DEFICIENCIES AND DISEASES OF THE VASCULAR SYSTEM

Probably the most common blood complaint is that of *anaemia* which means loss of normal balance between the productive and destructive blood processes. This can be due to a drop in the blood volume after a haemorrage, or a drop in the number of red blood cells, or in the amount of haemoglobin, or a combination of any two or more of these factors.

There are many forms of anaemia but we are primarily concerned with two categories, *simple anaemia* and *pernicious anaemia*.

In simple anaemia there are two direct causative factors, the first is a marked nutritional deficiency of iron, frequently seen in the premature infant, the

the growing child, and the pregnant woman. The second causative factor is chronic blood loss, for example during menstruation or because of accident.

One of the characteristics of pernicious anaemia is the presence of giant red cells (*macrocytes*), each cell appearing to be overloaded with haemoglobin, whilst the total red cells count is decreased. As recently as 1925 this disease was invariably fatal—today the life expectancy of the properly treated patient is about the same as that of the general population. Basically, pernicious anaemia results from failure of red blood cells to develop and mature normally.

Whilst a decreased number of red blood cells is indicative of anaemia a continuously increasing number of white blood cells can be indicative of *leukaemia*. Reference has already been made to the fact that white blood cells increase in number by mitosis in the presence of the necessary stimuli such as an infection and the normal 8000 per cubic millimetre of blood can increase to as many as 60 000 in a case of severe pneumonia. However, when the condition is cured, the mitosing or dividing ceases and the white blood count returns to normal. In leukaemia the leucocytes and/or lymphocytes do not remain at the normal number but gradually increase.

Varicose Veins

A network of veins serves to drain the capillary beds and body tissue of 'used' blood, and returns this blood to the heart. Venous flow is assisted in its return to the heart by the rhythmic suction action of breathing, muscular contraction in the extremities and the valves located in the veins. Gravity assists the venous blood from the neck and head to return to the heart but venous flow from the legs is against the pull of gravity and, for most of the day, has to run uphill. The valves in the veins prevent back flow and when some of these valves become impaired or cease to function the veins become permanently dilated. There are many causes of varicosity but these include:

(1) *Congenital factors*: varicosity appears to run in families.

(2) *Environmental factors*: people whose work necessitates their standing still for long periods of time are at particular risk.

Varicose veins are also, quite often, a complication of pregnancy and obesity.

Haemophilia

This is the best known of the bleeding diseases. It is a hereditary disease—the victim is always male and the disease is passed on by the mother, who is the so-called carrier. It is a disease in which there is a deficiency in the clotting of the blood.

Blue Baby

This is a baby born with a congenital structural defect of the heart which results in a constant recirculation of some of the venous blood without its prior passage through the lungs to pick up oxygen. The degree of blueness is, at least in part, dependent on the size of the hole through which the venous blood passes.

BLOOD TRANSFUSIONS

The transfer of blood to a recipient from a donor is one of the very widely used procedures in medical treatments—making up deficiencies caused by severe haemorrhage and, in some cases, when the blood volume is normal, a transfusion is used in order to replace a deficiency in one of the constituents of the blood.

The first record we have of a transfusion was of one performed between two dogs by a Richard Lower in England in 1665. Soon after this it was tried in France but the results on humans were so disastrous that the French passed a law forbidding transfusions. It was not until the early 20th century, when Karl Landsteiner completed his blood grouping, that progress was made in the field of human transfusions. Because, at that time, they had no means of keeping the blood fresh— only direct transfusions were possible. In 1914 Louis Agote of Argentina found that sodium citrate could be used for this purpose and the discovery was used extensively in the First World War. Since that time new methods have been found for obtaining and keeping blood for use at some future time and blood banks have become an accepted part of our medical system.

Arteriosclerosis and Atherosclerosis

These two conditions are often confused because of the similarity in many of the symptoms.

Simply, arteriosclerosis is hardening of the arterial walls brought about mainly by degenerative changes which increase in frequency with age. In athero-sclerosis—there is a build up of cholesterol on the inside of the artery which reduces the size of the bore.

GLOSSARY

Angiology — the science dealing with blood vessels and lymphatics

Cholesterol — a constituent of all animal fats and oils, insoluble in water. Its presence on the inside walls of blood vessels contributes to hypertension and other cardio-vascular conditions

Coronary — relating to the blood

Diastolic Pressure — the pressure measured during the relaxing phase of the cardiac cycle

Electrocardiogram (ECG) — a graphic record of heart activity made on an instrument known as an electro-cardiograph

Haemorrhoids (Piles) — dilated veins in the rectum and anus, described as internal or external depending on their position

Hypertension — high blood pressure

Hypotension — low blood pressure

Phlebitis — an inflammation of the vein walls, most common in the legs. It may lead to thrombo-phlebitis, a complication caused by an obstructing blood clot.

Systolic Pressure — the pressure measured during the contraction phase of the cardiac cycle

Thrombus — a clot of blood found within the heart or blood vessels

Tricuspid and *Mitral* — valves of the heart

Chapter 5

The Neurological System

The neurological system of the body has two main divisions—

(1) **The central** (or cerebrospinal) *nervous system.*
(2) **The autonomic** (including the sympathetic and parasympathetic) *nervous system.*

The basis of the nervous system is the *nerve cell* or *neuron*. This consists of a nerve cell body with its receiving processes—the *dendrites*, its transmitting process—the *axon* and its *nerve endings*. White nerve fibres are *medullated*, that is they are enclosed in a sheath of *myelin*. Grey nerve fibres are non-medullated, that is, they have no myelin.

A NEURON

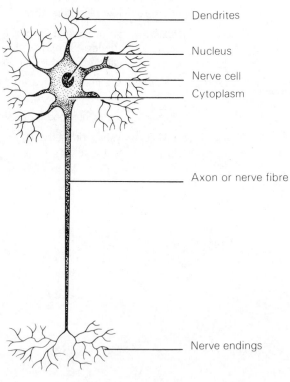

Dendrites

Nucleus

Nerve cell

Cytoplasm

Axon or nerve fibre

Nerve endings

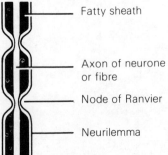

Fatty sheath

Axon of neurone or fibre

Node of Ranvier

Neurilemma

DIAGRAMMATIC VIEW OF A NERVE SHEATH

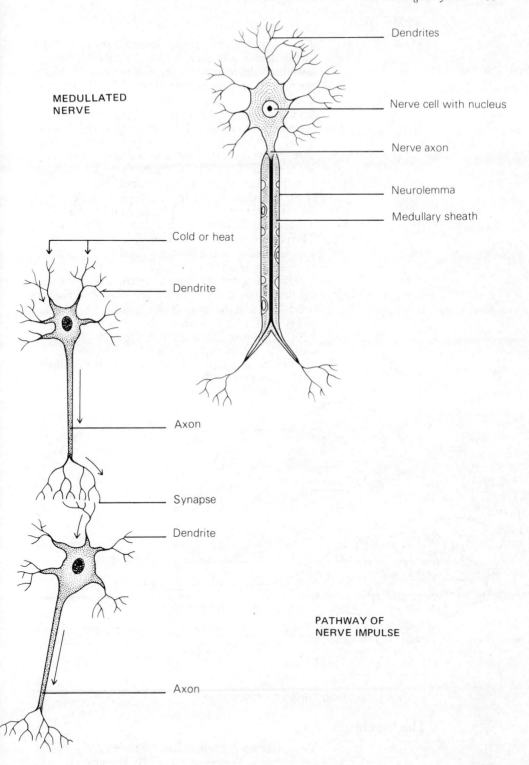

MEDULLATED
NERVE

Dendrites

Nerve cell with nucleus

Nerve axon

Neurolemma

Medullary sheath

Cold or heat

Dendrite

Axon

Synapse

Dendrite

Axon

PATHWAY OF
NERVE IMPULSE

THE BRAIN

At the centre of the nervous system is the brain. This, as has already been seen, is well protected from the outside by the hard bone structure of the skull. Inside, the brain is protected externally by three membranes known as the *meninges*. The outer layer is known as the *dura mater* (strong or hard mother), the middle layer is known as the *arachnoid* and the inner layer as the *pia mater* (soft mother). The outer meninges (the dura mater) is constructed of strong fibrous tissue anchored to the skull. The middle tissue, or arachnoid, is much more delicate and is not anchored to the skull, thus allowing the brain to expand. Under it lies the big reservoir of cerebral spinal fluid by which the whole of the brain is surrounded and on which it rests. Then comes the pia mater which is in contact with the grey matter of the brain itself and dips deep down between the brain convolutions.

When we speak of the brain we are really considering three quite different structures—the *cerebrum*, the *cerebellum* and the *medulla oblongata*.

The adult human brain weighs rather more than 1360 g and is so full of water that it tends to slump rather like a blancmange if placed without the support of a firm surface. It is estimated that it has 12 billion neurons or nerve cells.

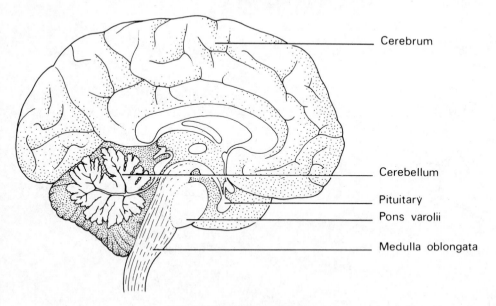

SECTION THROUGH THE BRAIN

Cerebrum

Cerebellum

Pituitary

Pons varolii

Medulla oblongata

The Cerebrum

The cerebrum consists of two symmetrical hemispheres. The outer layer of the cerebrum is known as the *cortex* and this is arranged in convolutions, that is, deep irregularly shaped fissures or indentations. This is the grey matter of the brain. Underneath the cortex lies nerve fibre or white matter. The function of the cerebrum is to control voluntary movement and to receive and interpret conscious sensations. It is the seat of the higher functions such as the senses, memory, reasoning, intelligence and moral sense.

The Cerebellum

The cerebellum is much smaller in size and lies below and behind the cerebrum. It too has grey matter under which is white matter. Its function is to control muscular co-ordination and balance.

The Medulla Oblongata

The medulla oblongata is about 3 cm long, tapering from its greatest width of 2 cm and connecting the rest of the brain with the spinal cord with which it is continuous. It is made up of interspersed white and grey matter. The medulla oblongata not only acts as the link between the brain and the central nervous system of the body but it is also the centre of those parts of the autonomic nervous system which control the heart, lungs, processes of digestion, etc.

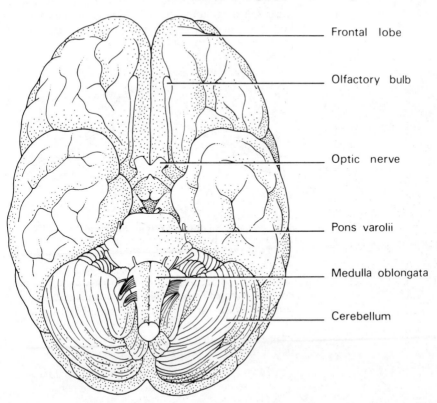

Frontal lobe

Olfactory bulb

Optic nerve

Pons varolii

Medulla oblongata

Cerebellum

UNDER SURFACE OF THE BRAIN

The *Pons Varoli*

Other parts of the brain include the *pons varoli* which is a bridge of nerve fibres linking the right and left hemispheres and also the cerebellum with the cerebrum above and the medulla oblongata below. All impulses which pass between the brain and the spinal cord traverse the pons varoli.

The Pituitary Gland (or Hypophysis)

The pituitary gland (or hypophysis) is a small gland about the size of a pea and lies in the pituitary fossa in the base of the skull. Its function is dealt with in the chapter on the endocrine system.

The Hypothalamus

The hypothalamus is situated in the area of the floor of the third ventricle of the brain and it exercises an influence over the autonomic nervous system. It contains the heat regulating centre and is generally believed to be involved with appetite.

The Spinal Cord

The spinal cord, which is continuous with the medulla oblongata, extends downwards through the vertebrae of the spinal column. The cord itself is cylindrical in shape with an outer covering of supporting cells and blood vessels and an inner egg-shaped core of nerve fibres. It extends through four-fifths of the spinal column and is about 45 cm in length.

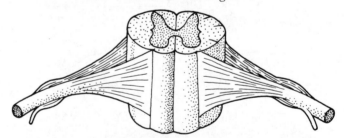

There are 12 pairs of cranial nerves given off from the base of the brain; 31 other pairs branch off the spinal cord throughout its length. These extend to every part of the body. Nerves that extend upwards through the spinal cord to the brain pass through the medulla oblongata where they cross—thus the left-hand side of the brain controls the right-hand side of the body whilst the right-hand side of the brain controls the left-hand side of the body. Nerves of the central nervous system fall into three categories:

(1) *Motor or efferent nerves*—the primary function of these nerves is to control the movement of muscles.

(2) *Sensory or afferent nerves*—these carry impulses from the sensory nerve endings to the spinal column and the brain.

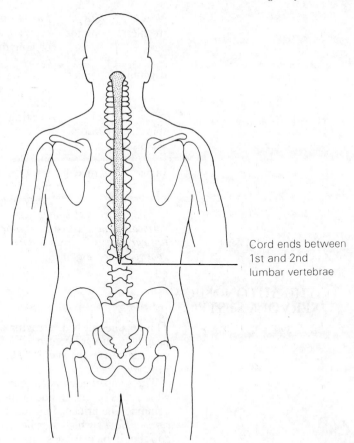

**POSITION OF
SPINAL CORD**

Cord ends between
1st and 2nd
lumbar vertebrae

(3) *Mixed nerves*—these consist of both motor and
sensory fibres.

Cranial Nerves

Name	Type	Function	Number
abducent	motor	supplies lateral rectus muscles of eyes	VI
auditory	sensory	sense of hearing, maintenance of balance, equilibrium	VIII
facial	mixed	sense of taste from tongue and impulses to muscles of facial expression	VII
glosso-pharyngeal	mixed	sensations from tongue, impulses to muscles of pharynx	IX
hypoglossal	motor	supplies tongue muscles	XII
oculomotor	motor	supplies muscles operating eyes	III
olfactory	sensory	sense of smell	I
optic	sensory	sense of sight	II

Name	Type	Function	Number
trochlear	motor	supplies superior oblique muscles of eyes	IV
trigeminal	mixed	receiving pain, heat, pressure and stimulating muscles of mastication	V
spinal accessory	motor	to sterno-cleido mastoid and trapezius muscles	XI
vagus	mixed	sensory, motor, digestive and respiratory organs	X

The 31 pairs of spinal nerves comprise 8 pairs of *cervical nerves*, 12 pairs of *thoracic nerves*, 5 pairs of *lumbar nerves*, 5 pairs of *sacral nerves* and 1 pair of *coccygeal nerves*.

THE AUTONOMIC NERVOUS SYSTEM

This supplies all body structures over which we have no voluntary control. It is divided into two separate parts—the *sympathetic system* and the *parasympathetic system*.

The sympathetic system comprises a gangliated cord which runs on either side of the front of the vertebral column. The principal plexuses of this system are: the *cardiac plexus* which supplies all the thoracic viscera and the thoracic vessels; the *coeliac* or *solar plexus* which supplies all the abdominal viscera and the *hypogastric plexus* which supplies the pelvic organs.

The parasympathetic nervous system consists mainly of the vagus nerve which gives off branches to the organs of the thorax and abdomen, but also includes branches from other cranial nerves, mainly the third, seventh and ninth as well as nerves in the sacral region of the spinal column.

It will be seen from the above notes that all the internal organs therefore have a double nerve supply from the sympathetic and parasympathetic systems and their effect is opposite—simply, a sympathetic nerve has the effect of increasing body activity and speeds it up, whereas the parasympathetic, on the contrary, slows down body activity.

The sympathetic fibres increase the heart rate, raise the blood pressure, mobilise glucose, stimulate the secretion of sweat. The parasympathetic fibres slow the heart, lower the blood pressure and decrease the secretion of sweat. It has been maintained that the sympathetic system provides for today's work and that its action increases when involved with physical activity. The parasympathetic—on the other hand—looks after tomorrow, being mainly concerned with changes which take place during rest.

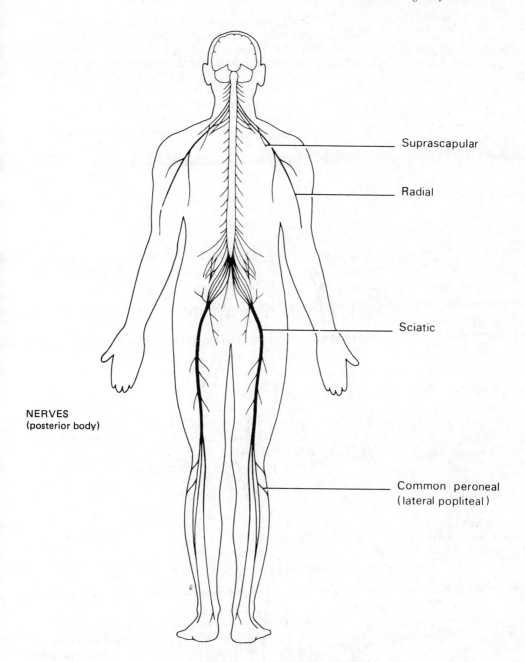

Suprascapular

Radial

Sciatic

Common peroneal
(lateral popliteal)

NERVES
(posterior body)

The sympathetic nerves are stimulated by strong emotions such as anger and excitement. In fact it is because of this effect of the emotions that they are called sympathetic.

The *adrenal* is one of the glands which they stimulate and the liberation of *adrenalin* is one of the body's responses to anger. In some people, the parasympathetic nerves are the stronger and hold the balance in the body; such people generally have a placid disposition, good digestion and are not very easily disturbed. These are known as *vagotonic types*. In other people, the sympathetic nerves are the stronger and these people are more emotional, less stable and their digestion is more readily disturbed. These are known as *sympatheticotonic types*.

Another function of the autonomic nervous system is related to the reflex nervous action. This is an involuntary reaction to a stimulation, for example,

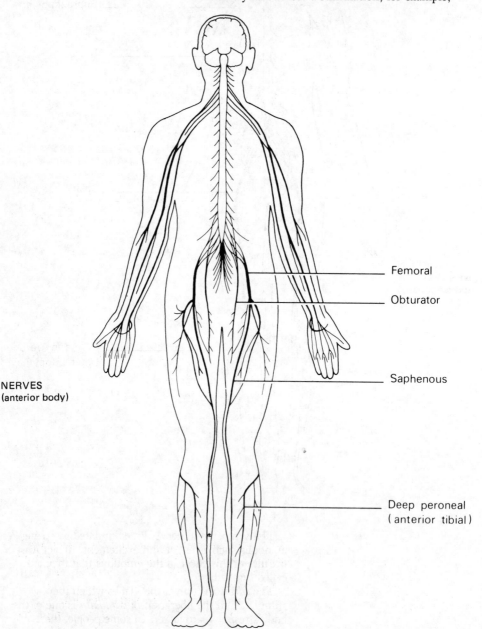

NERVES
(anterior body)

Femoral

Obturator

Saphenous

Deep peroneal
(anterior tibial)

taking the fingers away quickly from a hot surface, the recovery of balance to prevent a fall, and so on. It is also within this system that a reflex action is *conditioned*. For example, the normal reflex action when handed a very hot plate would be to drop it, but as this action would carry with it certain distinct disadvantages, like loss of the meal that was on the plate or the work involved with clearing up afterwards, the plate—instead of being dropped—is quickly put down. That is a reflex action which has been conditioned by other considerations.

CONDITIONS AND DISEASES OF THE NEUROLOGICAL SYSTEM

Neuritis

This takes in a wide group of disturbances which affect the peripheral nerves after they leave the spinal cord. Some of the disturbances are due to infection—others to compression of the nerves. Probably the biggest single factor is the build-up of urea and lactic acid at a point, or points, of the nerve's course, which affects the nerve's sheathing.

Bell's Palsy or Facial Paralysis

A neuritis of the facial nerve usually caused by infection and compression of the swollen nerve as it passes through a tiny opening in the skull below the ear in its course to the muscles of the face.

Neuralgia

This is a painful condition in a nerve due to irritation, inflammation or exposure.

Parkinson's Disease

Otherwise known as Paralysis Agitans, this is an extremely common illness beginning in middle life, deriving from disease of the basal ganglia. The disease is slowly progressive but as it does not affect the brain there is no loss of speech and intelligence is unaffected. The chief symptoms of this illness are tremor, rigidity and slowness of movement.

Sciatica

This is inflammation of the great sciatic nerve, the longest single nerve in the body. This is often a form of rheumatic neuritis but it can also be caused by compression, an arthritic spur or a slipped disc.

GLOSSARY

Brachial Neuritis	a condition similar to sciatica but in the arm
Ganglion	a group of nerve cell bodies usually located outside the brain and spinal cord
Plexus	a network of interlacing nerves
Spasticity	a state of sustained contraction of a muscle associated with an exaggeration of deep reflexes
Synapse	the region of communication between neurons; the point at which an impulse passes from an axon of one neuron to a dendrite of the cell body of another

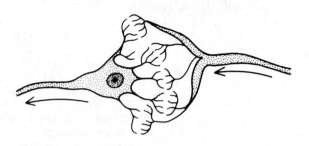

NERVE SYNAPSE

Chapter 6

The Digestive System

This is the system which is responsible for changing the food, which is *put into* the body, into substances suitable for absorption and therefore *usable by* the body. As the health and efficient working of the body must depend, to a very large extent, on the food which is put into it and the treatment which the food receives—it is necessary to have at least a basic

THE DIGESTIVE
SYSTEM

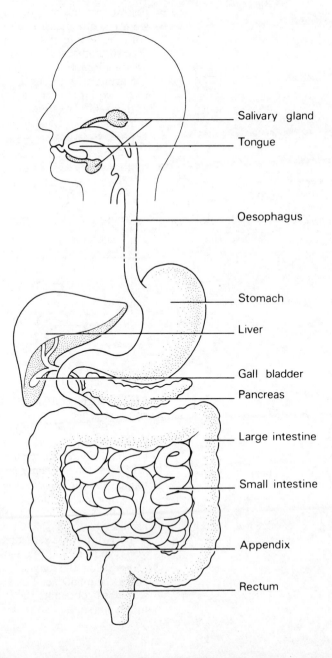

Salivary gland

Tongue

Oesophagus

Stomach

Liver

Gall bladder

Pancreas

Large intestine

Small intestine

Appendix

Rectum

understanding of the processes involved and some of the ways in which they may go wrong. The body needs material for growth, repair, heat and energy and these materials are supplied by the foods we eat. It is the digestive system which produces the chemical and other changes which make it possible for the food to perform functions necessary to maintain life.

Digestive juices contain *enzymes* which break down food. Enzymes are proteins which speed up chemical reactions; they are biological *catalysts*.

The digestive tract or alimentary canal is more than 10 m long. It is continuous, starting at the mouth, passing through the pharynx, the oesophagus, the stomach, the small and large intestines and ending with the rectum and the anus. Associated with it are accessory organs: the *tongue, teeth, salivary glands, liver* and *pancreas*.

Starting with the teeth—there are 32 permanent teeth; working from the front backwards on each side of the jaw, there are *two incisors, one canine* or eye tooth, *two premolars* or bicuspids and *three molars*.

THE TEETH AND BONY PALATE

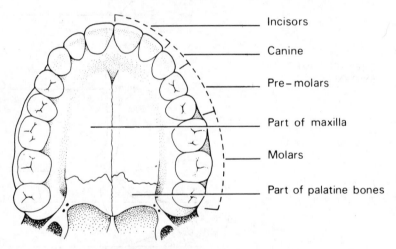

Incisors

Canine

Pre-molars

Part of maxilla

Molars

Part of palatine bones

The tongue consists of striated voluntary muscle and is attached mainly to the mandible and hyoid bones. The upper surface of the tongue is covered with papillae—there are three forms—the *filliform papillae* found chiefly on the dorsum of the tongue, the *fungiform* found mainly on the sides and tip of the tongue and the *vallate*—the largest of the papillae—lying in a V formation at the back of the tongue. Taste buds are resident in the walls of the vallate papillae.

There are three pairs of salivary glands, the *parotid glands* in front and below the ears, the *sublingual glands* below the tongue and the *submandibular glands* below the mandible. The salivary glands produce secretions containing the enzyme, *ptyalin,* which helps in the digestion of cooked starches.

From the mouth the food passes into the *pharynx* which is a muscular tube that has seven openings into it. These are the *mouth*, the *oesophagus*, the *larynx, two posterior apertures of the nose* and *two auditory (Eustachian) tubes from the ear*.

From the pharynx the food passes into the oesophagus which is a muscular tube lined with mucous membrane and covered with fibrous tissue. From here the food passes into the stomach which is a muscular sac, its size and shape varying with its contents and muscular tone. The stomach presents two curvatures, the *greater* and the *lesser curvature* and is divided into three parts—the *cardiac portion*, the *body* and the *pyloric*. The openings into the stomach are guarded by circular bands of muscle, the *cardiac sphincter* muscle at one end and the *pyloric sphincter* muscle at the other. The stomach has three coats or coverings, the outer coat of *serous membrane*, the *middle muscular coat* and the inner *mucous membrane*. This mucous membrane is arranged in folds or rugae which disappear when the stomach is distended. The membrane is lined with glands which produce gastric juice. This contains the enzymes *pepsin* (responsible for protein digestion) and *rennin* (responsible for the curdling of milk) and also hydrochloric acid.

THE STOMACH
(anterior view)

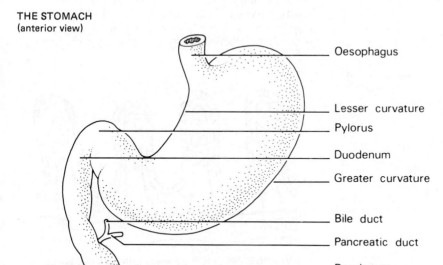

- Oesophagus
- Lesser curvature
- Pylorus
- Duodenum
- Greater curvature
- Bile duct
- Pancreatic duct
- Duodenum

From the stomach the food passes into the smaller intestine, the first part of this being the *duodenum* which is about 25 cm long and shaped like a letter 'C'. The remainder of the small intestine consists of the *jejunum* which is about 2.5 m long and the *ileum* which is about 3.5 m long. The inner coat of the small intestine is comprised of mucous membrane arranged in folds known as *valvulae conniventes* and—unlike the rugae of the stomach—these folds do not disappear with the distension of the intestines.

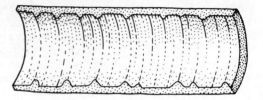

SECTION OF SMALL INTESTINE
(showing puckered lining of
valvulæ conniventes)

The mucous membrane is covered with minute fingerlike projections known as *villi*; each villus contains a lacteal for the absorption of fat and a capillary loop for the absorption of sugar and protein. This mucous membrane also contains intestinal glands which produce a secretion known as *succus entericus* which contains enzymes for the digestion of protein and sugars. The mucous membrane is studded with lymphatic nodules and, in the latter part of the small intestine, that is in the ileum, groups of these nodules are found and are known as *Peyer's patches*, their function being to fight infection. The small intestine then merges with the large intestine which though wider than the small intestine is much shorter, about 1.5 m long.

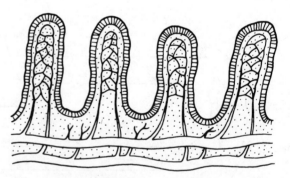

ENLARGED SECTION OF SMALL INTESTINE WALL
(showing villi)

The large intestine can be divided into nine parts; it starts with the *caecum* into which the ileum opens. The opening is guarded by the *ileo-caecal valve* which allows onflow but prevents backflow of intestinal contents. The *vermiform appendix* is attached to the blind end of the caecum and is about 7.5 cm long. The *ascending colon* passes upwards from the caecum along the right side of the abdomen and bends sharply

to the left at the *right* or *hepatic flexure* to become the
transverse colon. This passes across the abdominal
cavity and turns sharply downwards at the *left* or
splenic flexure to continue as the *descending colon*.
This goes down the left side of the abdomen to the
sigmoid colon in the pelvic cavity and the *rectum*. The
rectum is about 13 cm long with 2 sphincter muscles
at the exit or *anus*.

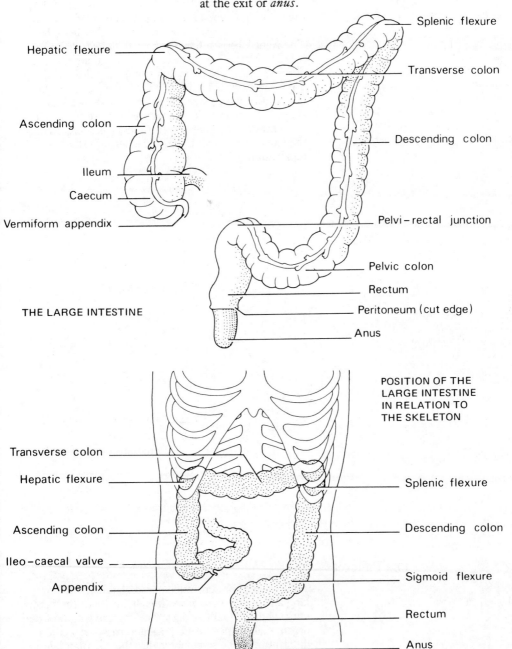

THE LARGE INTESTINE

POSITION OF THE
LARGE INTESTINE
IN RELATION TO
THE SKELETON

Having examined the alimentary canal it is necessary to look at the supporting organs of digestion.

The Liver

The liver is situated on the right hand side of the body just below the diaphragm. This is really a gland and is the largest organ in the body. It measures about 25–30 cm across and 15–18 cm from back to front; it weighs approximately 1.5 kg. It is divided into two lobes—the large right lobe and the smaller left lobe. The right lobe is subdivided into the *quadrate* and *caudate* lobes. The liver has many functions and one of these is the formation and storage of bile—of which it produces up to 1 litre in a day. This passes to the gall bladder which is a muscular, pear-shaped sac about 7.5 cm long. Its function is to store bile and to concentrate it by eight to ten times; when required, the bile passes out of the gall bladder into the duodenum.

THE GALL
BLADDER AND
ITS DUCTS

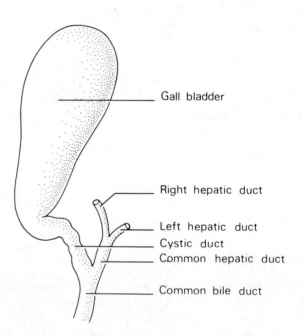

Gall bladder

Right hepatic duct

Left hepatic duct
Cystic duct
Common hepatic duct

Common bile duct

The Pancreas

The pancreas is a cream-coloured gland, 15–20 cm long and about 4 cm wide. It is divided into the head, neck, body and tail. A duct, running the length of the organ, collects pancreatic juice and passes it to the duodenum at the same point that the common bile duct passes in bile. The *islets of Langerhans* are specialised cells of the pancreas which produce *insulin*.

This is passed into the general circulation and controls carbohydrate metabolism.

ADDITIONAL NOTES ON THE PRINCIPLES OF DIGESTION

During the digestive process, large particles of protein, carbohydrate and fat are reduced in size and converted into simpler substances enabling them to be absorbed through the walls of the digestive tract into the blood stream.

Proteins are broken down to peptones and polypeptides and finally amino acids. Large particles of carbohydrates (starches or polysaccharides) are reduced to *disaccharides* which, in turn, are reduced to *monosaccharides*. Fats are split into their component parts, *fatty acids* and *glycerine*.

It should be noted that with one or two exceptions there is no absorption of food elements until they reach the intestine, where fatty acids and glycerine pass into the lacteals of the villi and amino acids into the capillary blood vessels. Fatty products are conveyed to the lymphatic system and enter the systemic circulation via the *thoracic duct*. Amino acids and simple sugars are carried by the portal vein to the liver.

The movement of food along the digestive tract is made possible by wavelike, muscular contractions known as *peristalsis*—the action is from the outside of the digestive tubes inwards and downwards, so that the food is forced further along the tube.

The stomach, being a muscularly controlled sac, is always on the move and might be compared with an old fashioned butter churn where the food is pushed around until it is well and truly mixed with gastric juice, a mixture of enzymes in hydrochloric acid.

As we have already seen, the stomach has a pyloric valve at the point where it merges into the small intestine. The function of this valve is to control the release of the partially digested food material into the small intestine. Watery foods, such as soup, leave the stomach quite quickly whilst fats remain considerably longer. An ordinary mixed diet meal is emptied from the stomach in 3–5 hours.

It has already been seen that the liver manufactures and stores bile but it has a variety of other functions. It is a powerful detoxifying organ, breaking down many kinds of toxic molecules and rendering them harmless. It is a reservoir for blood and a storage organ for some vitamins and digested carbohydrate in the form of glycogen, which it releases to sustain blood sugar levels. It manufactures enzymes, cholesterol, proteins, vitamin A from carotene, blood coagulation factors and other substances.

Bile is a complex fluid containing amongst other things bile salts and bile pigments. The pigments are derived from the disintegration of red blood cells and give the yellow brown colour of the faeces which are excreted. The bile salts are reabsorbed and reused; they promote efficient digestion of fats by a detergent action which gives very fine emulsification of fatty materials.

SOME CONDITIONS AND DISEASES OF THE DIGESTIVE SYSTEM

Appendicitis

This is an acute inflammation of the vermiform appendix. A distended, inflamed appendix may rupture—in which case it produces toxic materials and can cause peritonitis which is an acute inflammation of the abdomen.

Cirrhosis of the Liver

There are several types of cirrhosis of the liver but *portal cirrhosis* is, by far, the most common. This is also referred to as gin drinker's liver, or alcoholic liver. It is usually caused by exposure to poison, which can include such substances as carbon tetrachloride and phosphorus, but by far the most common cause is the ingestion of alcohol. This makes the liver leathery and produces nodules on its normally smooth surface—varying in size from a pin head to a bean— which give it a hobnailed appearance.

Jaundice

Jaundice is normally evidenced by the yellowness of the skin caused by an excess of bile pigments in the circulatory system. It may occur when the outflow of the bile has been blocked and when the liver surface itself is inflamed. When the small bile duct within the liver becomes obstructed a large portion of the bile which is produced by the liver is absorbed directly into the blood stream, as it cannot flow normally out of the bile duct into the duodenum.

Heartburn (Pyrosis)

This is a common symptom of gastric distress consisting of a burning sensation which extends up into the oesophagus and quite often into the throat and is accompanied by a sour belch.

Chapter 7

The Respiratory System

The respiratory system is responsible for taking in oxygen and giving off carbon dioxide and some water. It divides into the upper respiratory tract and the lower respiratory tract. The process of taking in air into the body is *inspiration* and getting rid of air from the body is *expiration*.

THE RESPIRATORY
SYSTEM

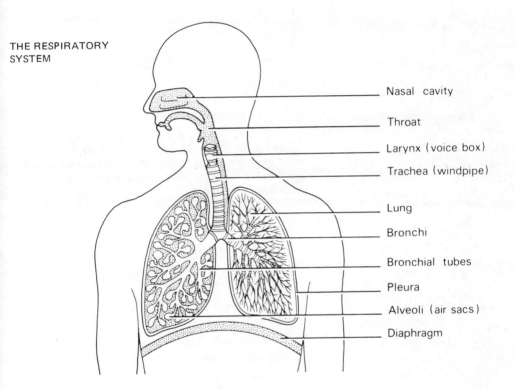

Nasal cavity

Throat

Larynx (voice box)

Trachea (windpipe)

Lung

Bronchi

Bronchial tubes

Pleura

Alveoli (air sacs)

Diaphragm

There are several organs involved in the respiratory system, the first being the nose. This is part of the *upper respiratory tract* which includes the mouth, the throat, the larynx and numerous sinus cavities in the head. Air brought in through the nose is filtered and warmed before passing down a tract into the lungs. The *lower respiratory tract* includes the trachea (or windpipe), the bronchi and the lungs, which contain bronchial tubes, bronchioles and alveoli, or air sacs.

The two lungs, which are the principal organs of the respiratory system, are situated in the upper part of the thoracic cage. They are inert organs, that is they do not work by themselves but function by a variation of atmospheric pressure which is achieved by a muscular wall known as the *diaphragm*.

**POSITION OF THE
LUNGS WITHIN
THE THORAX**

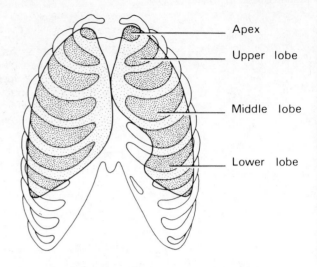

Apex

Upper lobe

Middle lobe

Lower lobe

The contraction and relaxation of the diaphragm
result in an alteration in the volume of the thorax, and
therefore an alteration of atmospheric pressure within
the lungs themselves.

The diaphragm when relaxed is a flattened dome

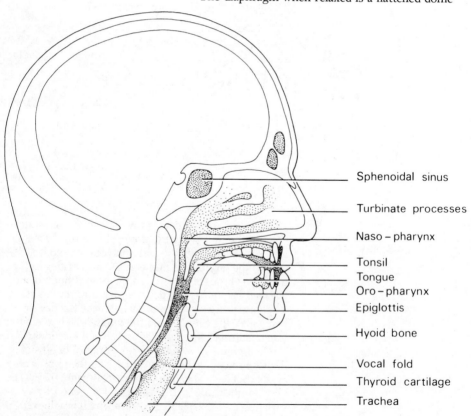

Sphenoidal sinus

Turbinate processes

Naso – pharynx

Tonsil
Tongue
Oro – pharynx
Epiglottis

Hyoid bone

Vocal fold

Thyroid cartilage

Trachea

THE UPPER RESPIRATORY PASSAGES

shape pointing upwards to the lungs. When it contracts, it flattens, pulls down the thorax, increases the volume of the thorax, and thus decreases the atmospheric pressure in the lungs. This causes air to rush in—inspiration. When the diaphragm relaxes, the thorax is pushed up, the volume decreases and the atmospheric pressure increases, and air rushes out of the lungs—expiration. The inspired air, which contains oxygen, passes down into the billions of minute air chambers or air cells known as *alveoli* which have very thin walls. Around these walls are the capillaries of the pulmonary system. It is at this point that the fresh air gives off its oxygen to the blood and takes carbon dioxide from the blood which is then expelled with the expired air.

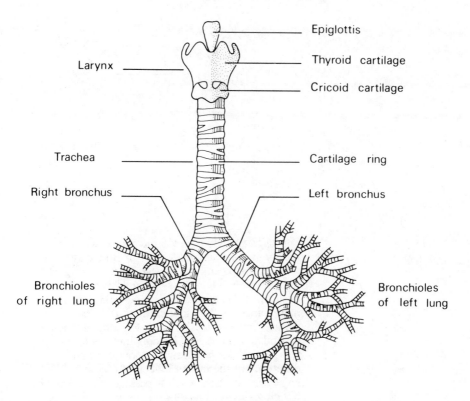

Epiglottis

Larynx

Thyroid cartilage

Cricoid cartilage

Trachea

Cartilage ring

Right bronchus

Left bronchus

Bronchioles of right lung

Bronchioles of left lung

THE RESPIRATORY PASSAGES

An average adult breathes something like 13 650 litres of air a day. This is not only the body's largest intake of any substance but also the most vital. It is possible to live without food for many days, without water for a few days but without air only for a very few minutes.

The trachea or windpipe measures about 11.5 cm in length and is approximately 2.5 cm in diameter. It has rings of cartilage to prevent it collapsing. It passes through the neck in front of the oesophagus branching into two bronchi—the right bronchus being 2.5 cm long and the left bronchus 5 cm long. The bronchi branch into smaller and smaller tubes ending in the bronchioles which have no cartilage in their walls and have clusters of the thin-walled air sacs—alveoli.

The *lungs* are greyish in colour and are spongy in appearance. The right lung has three lobes—upper, middle and lower, and the left lung has two lobes—upper and lower.

SECTION OF THE
LUNG

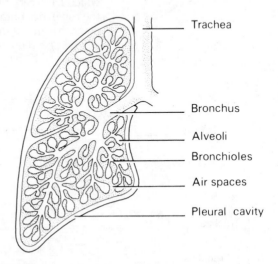

Trachea

Bronchus

Alveoli

Bronchioles

Air spaces

Pleural cavity

The *pleura* is the serous membrane which covers the lungs. The *visceral layer* is in close contact with the lung tissue and the *parietal layer* lines the chest wall. Between these layers is the *pleural cavity*. In health it is a natural cavity because the two membranes are fluid lubricated on their opposing surfaces and slide easily over each other as the lungs expand and contract. Air going into the lungs follows the same throat passageway as food for a short distance, but there is an ingenious trapdoor called the *epiglottis* which permits the passage of air to the lungs but closes it when food or liquids are swallowed.

During normal quiet breathing about 0.5 litres of air flows in and out of the lungs. This is known as *tidal air*. If inhalation is continued at the end of ordinary breathing, an additional 1.5 litres of *complemental air* can be forced into the lungs. If exhalation is continued, an extra 1.5 litres of *supplemental air* can be forced out of the lungs. About 1 litre of air remains in the lungs and cannot be expelled—this is known as *residual air*.

The normal rate of inspiration and expiration, the respiration rate, is about 16 times a minute in an adult.

CONDITIONS AND DISEASES OF THE RESPIRATORY SYSTEM

Bronchitis

This occurs in two forms—acute bronchitis and chronic bronchitis. Acute bronchitis may result from inhaled materials—fog, smoke, chemicals, etc., or it may be connected with another disease condition—influenza, measles or whooping cough. Chronic bronchitis normally occurs at middle age or later and is four times more prevalent in men than in women. The disease can prove fatal and about 30 000 deaths are recorded annually in Britain from this cause.

Pleurisy

Practically any disease that causes inflammation of the lungs may result in pleurisy. The pleura becomes inflamed and fluids accumulate in the interspace.

Pneumoconiosis

The term pneumoconiosis indicates a lung condition due to inflammation by minute particles of mineral dusts. It is often called miners' disease or miners' lung due to its prevalence amongst this body of workers. There are, however, a number of other occupations where fine dust is a hazard and workers who are exposed to a high concentration of silica dust may develop a variation of the disease known as *pneumosilicosis*. Precautionary measures like the wearing of masks help to reduce the incidence of this disease.

Pulmonary Tuberculosis

The disease we know as tuberculosis has been with us for thousands of years. Centuries before Christ it was called *Phthisis*—this is a Greek word meaning wastage or decay. This explains the familiar word for this disease—consumption. It was in 1882 that a German bacteriologist, Robert Koch, discovered that tuberculosis was caused by a long, thin bacterium called *tubercle bacillus*.

GLOSSARY

Asthma a paroxysmal condition usually due to hypersensitiveness to inhaled or ingested substances, e.g. pollen asthma.

Pneumothorax collapsed lung—may occur in accidents from bones perforating the chest. Artificial pneumothorax is the introduction of air or other gas into the pleural cavity through a needle in order to produce collapse and immobility of the lung. It is used in the treatment of pulmonary tuberculosis.

Rhinitis inflammation of the nasal mucous membrane, e.g. acute serous rhinitis—hay fever

Chapter 8

The Genito-Urinary System

In many anatomical textbooks this system is dealt with as two systems—the reproductive system and the excretive system. However, as a number of the organs involved are common to both systems the general tendency is to treat them under one heading.

The principal organs involved in the dual system are the ovaries, fallopian tubes, uterus, testes, urethra, ureter and the urinary bladder. There are a number of smaller accessory organs involved and these will be dealt with appropriately in the text.

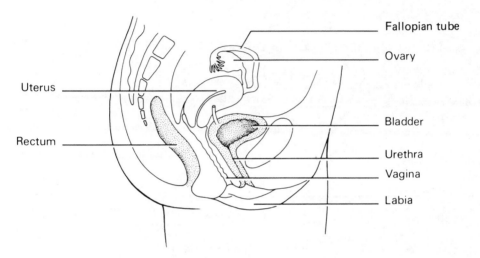

SECTION OF THE FEMALE PELVIC CAVITY

First we have the right and left *ovaries* in the female anatomy. These are quite small, about the size of an almond; they consist of masses of very small sacs known as the *ovarian follicles* and each follicle contains an egg—*ovum*. The ovaries have two principal functions:

(1) to develop the ova and expel one at approximately 28 day intervals during the reproductive life, and

(2) to produce hormones (oestrogen and progesterone) which influence secondary sex characteristics and control changes in the uterus during the menstrual cycle.

The *fallopian tubes*—sometimes referred to as the uterine tubes—are about 10 cm long and their function is to transport the ova from the ovaries to the uterus.

The *uterus* is a muscular organ approximately pear-shaped, about 7.5 cm long by 5 cm wide and 2.5 cm thick. It is positioned in the centre of the pelvis with the bladder in front and the rectum behind. It is normally divided into three parts—the *fundus*, the broad upper end, the *body*, the central part, and the *cervix* (about 2.5 cm long), the neck which projects into the vagina.

The *vagina* is the muscular canal which connects the above organs to the external body at the point collectively known as the *vulva* which includes the *clitoris*—a small, sensitive organ containing erectile tissue corresponding to the male penis.

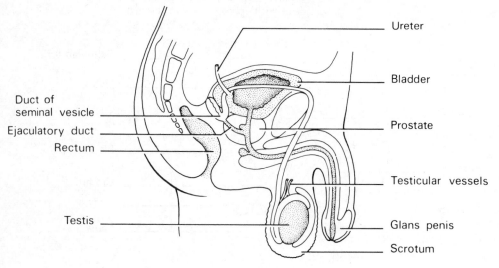

Ureter

Bladder

Duct of
seminal vesicle

Ejaculatory duct

Rectum

Prostate

Testicular vessels

Testis

Glans penis

Scrotum

MALE GENITAL ORGANS

The male genital organs are fairly simple in comparison to the female genital organs. The principal organs are the *testes* or testicles which are the essential male reproductive glands, the *scrotum* which is a pouch-like organ containing the testes, and the *penis* which is suspended in front of the scrotum.

The *kidneys* are two bean-shaped organs, approximately 10 cm long, 5 cm wide and 2.5 cm thick. They are positioned against the posterior abdominal wall at the normal waistline, with the right kidney slightly lower than the left.

The kidneys consist of three principal parts—the *cortex* or outer layer which is bright reddish-brown in colour, the *middle portion* or *medulla* which is inside and red striated in colour and the *pelvis* which is the hollow, inner portion from which the ureters open.

The function of the kidneys is to separate certain waste products from the blood and this renal function helps maintain the blood at a constant level of composition despite the great variation in diet and fluid intake. As blood circulates in the kidneys a large

SECTION OF
KIDNEY

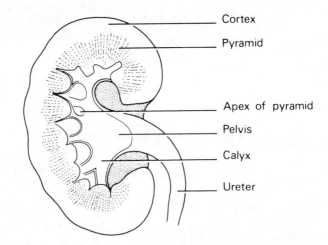

Cortex

Pyramid

Apex of pyramid

Pelvis

Calyx

Ureter

quantity of water, salts, urea and glucose is filtered into the *capsules of Bowman* and from there into the *convoluted tubules*. From here all the glucose, most of the water and salts and some of the urea are returned to the blood vessels—the remainder passes via the *calyces* into the kidney pelvis as urine. It is estimated that 150–180 litres of fluid are processed by the kidneys each day but only about 1.5 litres of this leaves the body as urine.

INTERNAL
KIDNEY
STRUCTURE

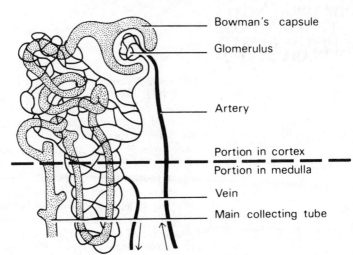

Bowman's capsule

Glomerulus

Artery

Portion in cortex
Portion in medulla

Vein

Main collecting tube

The *ureters* are two fine muscular tubes, 26–30 cm long, which carry the urine from the kidney pelvis to the bladder. This is a very elastic muscular sac lying immediately behind the *symphysis pubis*.

The *urethra* is a narrow muscular tube passing from the bladder to the exterior of the body. The female urethra is 4 cm long and the male urethra 20 cm long.

In the male, the urethra is the common passage for both urine and the semen or reproductive fluid. Also, in the male, it passes through a gland known as the *prostate gland* which is about the size and shape of a chestnut. It surrounds the neck of the bladder and tends to enlarge after middle life when it may—by projecting into the bladder—produce urine retention.

THE EXCRETORY SYSTEM

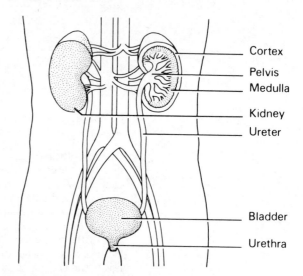

Cortex

Pelvis

Medulla

Kidney

Ureter

Bladder

Urethra

ADDITIONAL NOTES ON THE FUNCTIONS OF THE GENITO-URINARY SYSTEM

The average composition of urine is 96% water and 4% solid—2% urea and 2% salts. The 2% urea compares with 0.04% urea in blood plasma so it will be seen that concentration has been increased some fifty times by the work of the kidneys. The salts consist mostly of sodium chloride, phosphates and sulphates produced partly from the presence of these salts in protein foods. These salts have either to be reabsorbed or got rid of by the kidneys in sufficient quantities to keep the normal blood balance.

The urine also contains traces of a number of other substances, all of which combine to produce in the urine a reasonable pattern of the state of the body itself. Its analysis indicates a number of physiological states including the amount of alcohol in the body, whether a female is pregnant or not and whether a person has diabetes.

It is estimated that, at birth, there are some 30 000 ova or eggs in a female child. No fresh ova are formed after birth but—during the reproductive female life—that is, commencing between 10 and 16 years of age and concluding between 45 and 55 years of age, these ova develop within the follicles or sacs in which they are embedded. They come progressively nearer to the surface of the ovary where they mature and increase in size. About every 28 days, one of these follicles bursts and the ovum it contains, together with the fluid surrounding it, is expelled into the fallopian tubes and thence into the uterus where it may or may not be fertilised. If the ovum is fertilised by a male reproductive cell or *spermatozoon* it then attaches itself to the uterine wall and develops there. If the ovum does not become fertilised within a few days it is cast off and the process termed *menstruation* is initiated.

The spermatozoa which are responsible for fertilisation are contained in a substance known as *seminal fluid*. An average ejection of seminal fluid contains several hundred million of these mobile sperm which look rather like miniature elongated tadpoles, about 0.05 mm in length. Each one consists of a headpiece, a middle piece and a long whiplike tail piece. It is this vigorous tail piece or lashing tail which gives the spermatozoon its mobility. The single fertilized ovum soon becomes many cells which develop in a bag of membranes and soon fill the uterine cavity. At one part of this sac—the point where the ovum first embedded itelf in the uterine wall, the *placenta* or afterbirth develops. The umbilical cord contains blood vessels and runs from the navel of the foetus to the placenta. The placenta receives the mother's blood from the wall of the uterus and the infant's blood via the umbilical cord so that, at no stage, does the mother's blood pass directly into the child. It is through the placenta that the child's blood is able to absorb food, oxygen and water from the mother and, in turn, give off its waste products.

The skin is, of course, an organ very closely connected with the excretal system but—as it is a multipurpose organ—it is dealt with in the final chapter of this section of the book.

CONDITIONS AND DISEASES OF THE GENITO-URINARY SYSTEM

Cervicitis

This is an infection of the *cervix*, that is, the neck of the uterus, and is reasonably common. It may be due to gonorrhoea, syphilis or a specific infection.

Cystitis

Cystitis is inflammation of the bladder, a condition especially common in women. This is due to the fact that the urethra in women is very short and is a pathway to invasion by infecting organisms.

Kidney Stones

Stones in the kidney are quite common and precipitate

out of the urine, which is a complex solution of many substances. Surgical operations for the removal of stones have a very long history. The Greek doctor—Hippocrates—admonished fellow physicians not to cut out stones but to leave it to the specialist. Nearer to our time, Samuel Pepys describes his own operation for the 'cutting of stone' on March 26, 1658. He notes in his diary that he spent twenty-four shillings 'for a case to keep my stone that I was cut of'.

Nephritis or Bright's Disease

This was first described by Dr Richard Bright of London in 1827. The single disease which he diagnosed has now been subdivided into a number of conditions which may, in a broader way, be called nephritis—an inflammation of the kidney not resulting from infection in the kidney.

GLOSSARY

Calculus	a stone e.g. renal calculus—stone in the kidney
Catheter	a hollow tube which is placed into a cavity through a narrow canal to discharge fluid from the cavity, e.g. draining urine from the bladder for relief of urinary retention
Dysmenorrhea	painful menstruation
Ectopic Gestation	development of the embryo in the fallopian tube instead of the uterus
Enuresis	involuntary discharge of urine
Foetus	the unborn child dating from the end of the third month until birth
Intra-uterine	within the uterus; relating to conditions which occurred before birth
Menopause	also called *climacteric*; the physiological cessation of menstruation
Micturition	the act of passing urine
Parturition	the act of giving birth

Hormones are chemicals which cause certain changes in particular parts of the body. Their effects are slower and more general than nerve action. They can control long-term changes such as rate of growth, rate of activity and sexual maturity.

The *endocrine* or *ductless glands* secrete their hormones directly into the blood stream. The hormones are circulated all over the body and reach their target organ via the blood stream. When hormones pass through the liver, they are converted into relatively inactive compounds which are excreted by the kidneys. Tests on such hormonal end products in urine can be used to detect pregnancy.

The *endocrine system* consists of a series of glands which secrete hormones; they are found throughout the body and include the pituitary, thyroid, parathyroids, thymus, supra-renal or adrenal glands, part of the pancreas and parts of the ovaries and testes.

Although these glands are separate—it is certain that they are functionally closely related because the health of the body is dependent upon the correctly balanced output from the various glands that form this system.

The Pituitary Gland (Hypophysis)

This gland has been described as the leader of the endocrine orchestra. It consists of two lobes, anterior and posterior. The anterior lobe secretes many hormones, including the growth-promoting *somatotropic* hormone which controls the bones and muscles and in this way determines the overall size of the individual. Oversecretion of the hormone in children produces gigantism and undersecretion produces dwarfism. The anterior lobe also produces *gonadotropic* hormones for both male and female gonad activity. *Thyrotropic* hormones regulate the thyroid and *adrenocorticotropic* hormones regulate the adrenal cortex. It also produces *metabolic* hormones.

The posterior lobe produces two hormones—*oxytocin* and *vasopressin*. Oxytocin causes the uterine muscles to contract; it also causes the ducts of the mammary glands to contract and, in this way, helps to express the milk which the gland has secreted into the ducts. Vasopressin is an antidiuretic hormone which has a direct effect on the tubules of the kidneys and increases the amount of fluid they absorb so that less urine is excreted. It also contracts blood vessels in the heart and lungs and so raises the blood pressure. It is not certain whether these two hormones are actually manufactured in the posterior lobe or whether they are produced in the hypothalamus and passed down the stalk of the pituitary gland to be stored in the posterior lobe and liberated from there into the circulation.

The Thyroid

The right and left lobes of this gland lie on either side of the *trachea* united by the *isthmus*. Average size of each lobe is 4 cm long and 2 cm across but these sizes may vary considerably. The secretion of this gland is *thyroxine* and *tri-iodothyronine*. Thyroxine controls the general metabolism. Both hormones contain iodine but thyronine is more active than thyroxin. Under-secretion of this hormone in children produces cretinism; the children show stunted growth (dwarfism) and fail to develop mentally.

GLANDS OF THE BODY

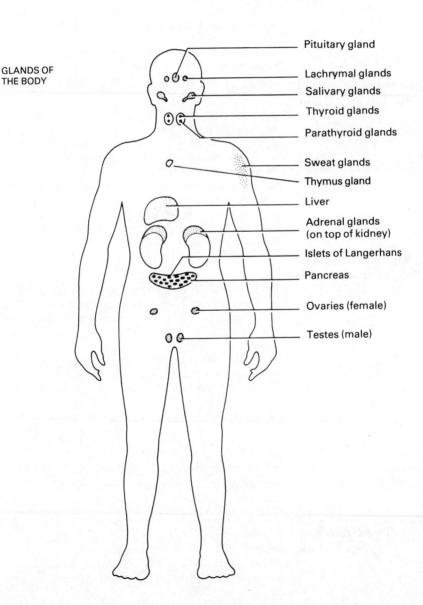

Pituitary gland

Lachrymal glands

Salivary glands

Thyroid glands

Parathyroid glands

Sweat glands

Thymus gland

Liver

Adrenal glands (on top of kidney)

Islets of Langerhans

Pancreas

Ovaries (female)

Testes (male)

Undersecretion in adults results in a low metabolic rate. Oversecretion in adults gives rise to *exophthalmic goitre* and the metabolic rate is higher than usual. Such persons may eat well but burn up so much fuel that they remain thin. This is usually accompanied by a rapid pulse rate. This gland, therefore, has a profound influence on both mental and physical activity.

The Parathyroid Glands

There are four of these glands, two on either side lying behind the thyroid. Their secretion is *parathormone*— the function of which is to raise the blood calcium as well as maintain the balance of calcium and phosphorus in both the blood and bone structures. Undersecretion gives rise to a condition known as *tetany* in which the muscles go into spasm, and oversecretion causes calcium to be lost to the blood from the bones giving rise to softened bones, raised blood calcium and a marked depression of the nervous system.

The Thymus Gland

This gland lies in the lower part of the neck and attains a maximum length of about 6 cm. After puberty the thymus begins to atrophy so that in the adult only fibrous remnants are found. Its secretion is thought to act as a brake on the development of sex organs so that as the thymus atrophies, the sex organs develop. Recent research into the activity of this gland reveals that it plays an important part in the body's immune system by producing T lymphocy—the T standing for thymus derived.

The Suprarenal or Adrenal Glands

These are two in number, triangular in shape and yellow in colour. They lie one over each kidney. They are divided like the kidney into two parts—the *cortex* and the *medulla*. The cortex is the outer part of the gland and produces a number of hormones called *cortico-steroids*. Their function is to control sodium and potassium balance, stimulate the storage of glucose and affect or supplement the production of sex hormones. The medulla or inner layer produces *adrenalin*, a powerful vasoconstrictor. Adrenalin raises the blood pressure by constriction of smaller blood vessels and raises the blood sugar by increasing the output of sugar from the liver. The amount of adrenalin secreted is increased considerably by excitement, fear, or anger, which has caused the adrenals sometimes to be referred to as the glands of fright and fight.

The Gonads or Sex Glands

These glands are naturally different in men and women because they serve different, though, in many respects, complementary functions. In the female the gonads are the ovaries and in the male the testes. Female sex hormones are *oestrogen* and *progesterone*. The male sex hormone is *testosterone*, though each sex produces a small quantity of the opposite hormone. The female hormones are responsible for developing the rounded, feminine figure, breast growth, pubic and axillary hair and all the normal manifestations of femininity and reproduction. Male hormone is responsible for voice changes, increased muscle mass, development of hair on the body and face and the usual development of manliness.

Pancreas

The endocrine part of the pancreas consists of clumps of cells called *islets of Langerhans* which secrete *insulin*. Insulin regulates the sugar level in the blood and the conversion of sugar into heat and energy. Too little insulin results in a disease known as *diabetes mellitus*. This disease is divided into one form, juvenile onset, which occurs before the age of 25, and another form which begins in maturity. It is a very common disease. It is known that some half million people in the United Kingdom suffer from it sufficiently badly to need treatment but it has been estimated that there are many more people in whom the disease exists at a sub-treatment level. Drs. Rankin and Best succeeded in 1922 in keeping a diabetic dog alive in their Canadian laboratory by injection of insulin. More recently, with surgery, it has been possible to contain this disease although the supplement of insulin is really a support treatment rather than a cure.

ADDITIONAL NOTES ON THE HORMONE SYSTEM

The quantities involved in the secretion of the various glands are minute. For example, the adrenal glands, which affect all the organs of the body, produce in a complete year not more than 1 g of hormone. Some hormone deficiencies appear to be endemic, that is they are particularly prevalent in certain parts of the world. The best example is probably to be found in the diseases which affect the thyroid through lack of iodine. For example, endemic cretinism is common to the upper valleys of the Alps and the Himalayas where endemic goitre is also present, whilst the latter condition is to be found also in the region of the Great Lakes and the Valley of Saint Lawrence. In all these areas, which are deficient in iodine or iodine-containing foods, the authorities now usually take precautionary measures such as the provision of iodised salts in order to stop the development of the disease.

GLOSSARY

Addison's Syndrome	a condition due to adrenal cortical tissue insufficiency, characterised by hypotension, wasting, vomiting and muscular weakness
Amenorrhoea	absence of menstruation
Cushing's Syndrome	condition due to oversecretion of adreno-cortical hormones, characterised by moon face, redistribution of body fat, hypertension, muscular weakness and occasionally mental derangement
Hyperthyroidism	thyrotoxicosis, toxic goitre and Graves' disease or exophthalmic goitre: the body's physical activities are subject to a speeding up whilst the opposite condition —hypothyroidism—is evidenced by a slowing down of the body's activities
Lesion	an alteration of structure or of functional capacity due to injury or disease
Menopause (Climacteric)	cessation of menses; the period in female development when the reproductive function comes to an end, linked to a decline in the supply of hormone secretions by the ovaries. This is not a disease—it marks another stage in the female progression through life
Premenstrual Tension	a syndrome of depression, irritability, bloating, swelling and restlessness that occurs for about one week before the onset of menstruation
Steroid	the generic name given to various compounds of internal secretions including the sex hormones
Syndrome	a group of symptoms and signs which, when considered together, characterise a disease

Chapter 10   Accessory Organs

The subject title of this chapter describes those parts of the human anatomy which do not fit completely into any one system although they may form part of it.

THE SKIN

The skin is an important organ which has three primary functions:

(1) It serves as a protective cover.
(2) It regulates body temperature.
(3) It provides a sensory covering over the entire body.

It has two principal divisions, the epidermis or outer skin and the dermis or true skin.

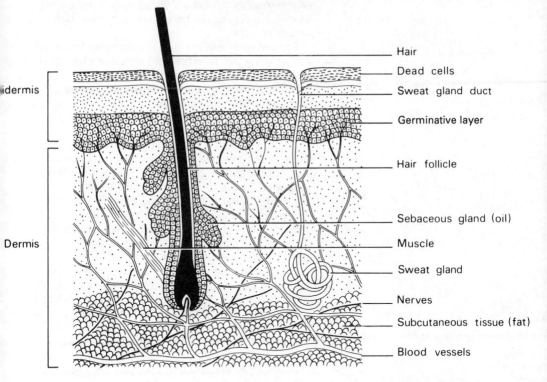

THE SKIN

Skin is a large organ. The average human adult is covered by about 1.7 m² of skin, varying in thickness from thin over the eyelids to thick on the soles of the feet. It weighs about 3.2 kg and provides an excellent protection against germs as very few can penetrate unbroken skin. Normal body processes produce heat and most of this is eliminated from the skin by radiation to the surrounding air or by evaporation of perspiration.

As an indication of the complexity of the skin it has been estimated that 1 cm^2 of skin contains approximately 3 million cells, 13 oil glands, 9 hairs, 100 sweat glands, 2.75 m of nerves, 1 m of blood vessels and thousands of sensory cells. It is, therefore, easy to see how healthy skin is indicative of good mental and physical health.

The *epidermis* is the outer layer of the skin which contains nerve endings with blood vessels to the germinative layer. It is nourished by tissue fluid derived from the dermis.

The *dermis* is a thicker layer of connective tissue which supports the hairs, each hair growing from a hair follicle.

The *erector pili* muscles, which are attached to the hair follicles, contract in response to cold and fear.

The *sebaceous glands* secrete sebum; *sweat* or *sudoriferous glands* extract water, salts, urea and other waste products and discharge them on to the skin surface as sweat.

Sensory nerve endings give sensations of touch, pain and temperature; superficial blood vessels play a part in regulating body temperature.

Nails are really appendages of the skin, being outgrowths from the epidermis.

Adipose tissue beneath the skin is one of the principal fat deposits of the body.

The *sweat glands* of the skin are of two types. The first produce *apocrine sweat* which has more social than physiological significance. The apocrine sweat glands are limited to a few regions of the body, primarily the axillary and genital areas. They are inactive in infants, develop with puberty and enlarge premenstrually. Freshly produced sweat is normally sterile and inoffensive but its decomposition by bacteria gives rise to perspiration odour.

The second kind of sweat is called *eccrine*—there are millions of these sweat glands all over the body and the sweat they give out is little more than diluted salt water. They are involved in the vital heat regulating system that enables the body to keep its constant internal temperature of 36.8 °C. These eccrine sweat glands disperse large quantities of water which, in extreme circumstances, can reach as much as 2.3 litres a day.

THE EYES

Our sense of sight is the response of the brain to light stimuli which are received through the eye. The eyeball is a hollow, spherical structure, its walls consisting of three principal layers:

(1) The *sclera* is a tough fibrous, opaque coat, which is modified in front to form the clear, transparent *cornea*.

(2) The *choroid* or middle coat consists of an interlacement of blood vessels and pigment granules supported by loose connective tissue; the *iris* is

a pigmented, muscular curtain suspended behind the cornea. In the centre of the iris is an aperture known as the *pupil* through which light reaches the interior of the eye.

(3) The *retina* forms the delicate inner layer of the eyeball. In this layer are found the *receptor* and *sensory* optic nerve endings sometimes referred to as *rods* and *cones*.

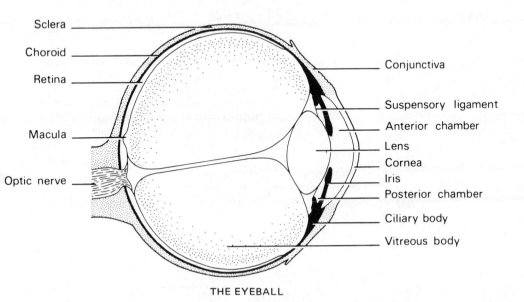

Sclera

Choroid

Retina

Macula

Optic nerve

Conjunctiva

Suspensory ligament

Anterior chamber

Lens

Cornea

Iris

Posterior chamber

Ciliary body

Vitreous body

THE EYEBALL

The eyeball has a number of appendages; the various muscles which directionally rotate it and the *lachrymal* or *tear glands* which moisten and clean the outer surface of the eye. Excess secretion of the lachrymal glands overflows onto the cheeks as tears. From the inner corners of the eyes the tears drain into a channel which opens into the nose, which is why weeping is sometimes accompanied by sniffing.

The pupil controls the light image by contracting in bright light and dilating in dim light. These light images strike the retina as an upside down image which is then conveyed to the brain through the optic nerve. The brain then reinverts the impulse so that it comes a right side up image.

THE EARS

The ear is made up of three parts—the *external* ear, the *middle* ear and the *internal* ear.

(1) **The external ear** consists of the *auricle* attached to the side of the head and the *external auditory meatus* leading from the auricle (or *pinna*) to the *tympanic membrane* or ear-drum. The function of the auricle is to collect sound waves and conduct them to the external auditory canal and tympanic

membrane. The external auditory meatus or canal contains ceruminous glands which secrete cerumen or wax.

(2) **The middle ear** or *tympanic cavity* is a small air-filled cavity containing a chain of small bones—*auditory ossicles*. Sound waves are transmitted from the tympanic membrane (parchment-like) by the auditory bones (malleus, incus and stapes, popularly known as the hammer, anvil and stirrup) to the oval window (*fenestra ovalis*), a membrane connecting with the internal ear.

The *Eustachian (auditory) tube* links the ear with the nasopharynx to ensure that air pressure in the middle ear is the same as atmospheric pressure. The middle ear also communicates with the *mastoid antrum* and mastoid air cells in the mastoid process of the temporal bone.

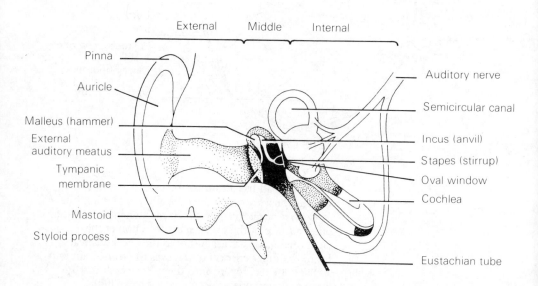

A SECTION THROUGH THE EAR

(3) **The internal ear** or *labyrinth* consists of bony cavities (*osseous labyrinth*) enclosing a membranous structure (*membranous labyrinth*) which approximately follows the shape of the bony labyrinth. Between the bony walls and the membranous part of the labyrinth is a clear fluid —*perilymph*. This transmits the vibrations from the oval window to the *cochlea* (the essential organ of hearing) which connects to the brain via the auditory nerve. The *three semi-circular* canals (membranous canals or ducts) are situated in the bony labyrinth and control balance.

THE MAMMARY GLANDS OR BREASTS

These are accessories to the female reproductive organs and secrete milk during the period of lactation. They enlarge at puberty, increase in size during pregnancy and atrophy in old age. The breast consists of mammary gland substance or *alveolar* tissue arranged in lobes and separated by connective and fatty tissues. Each lobule consists of a cluster of alveoli opening into lactiferous ducts which unite with other ducts to form large ducts which terminate in the excretory ducts. The ducts near the nipple expand to create reservoirs for the milk—*lactiferous sinuses*.

The breast contains a considerable quantity of fat, which lies in the tissue of the breast and also in between the lobes. It contains numerous lymphatic vessels which commence as tiny plexuses, unite to form larger vessels and eventually pass mainly to the lymph node in the axilla. The nipple is surrounded by a darker coloured area known as the mammary areola.

The breasts are greatly influenced by hormone activity. Hyper-secretion of the thyroid can lead to atrophy of the breasts whilst hypo-secretion can cause greatly developed breasts. Both the ovarian hormones influence the condition and appearance of the breast whilst the pituitary hormone, *prolactin*, starts lactation at the end of pregnancy.

A SECTION THROUGH THE BREAST

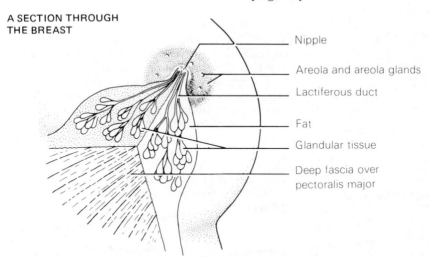

Nipple

Areola and areola glands

Lactiferous duct

Fat

Glandular tissue

Deep fascia over pectoralis major

CONDITIONS AND DISEASES OF THE ACCESSORY ORGANS

Acne Vulgaris or Common Acne

This is one of the most common forms of skin disease. It is due to an oversecretion of the sebaceous glands and is invariably associated with the increase of sex hormones at puberty in both male and female. The primary lesion of acne is the *comedo* or *blackhead*, the blackhead being darkened by air and not by dirt as sometimes thought.

Alopecia Areata (patchy baldness)

In this condition hair is lost from a small area, generally on the scalp but sometimes in the hirsute area of the face. There is no inflammation or any obvious skin disorder or systemic disease.

Cataract

The lens of the eye is situated directly behind the pupil and in health is clear. With age and in some diseases such as diabetes it loses its transparency and becomes more opaque, gradually shutting out vision —this condition is known as a cataract. It is not a growth but a biochemical change in the lens.

Conjunctivitis

This is inflammation of the conjunctiva. In its acute contagious form it is known as 'Pink Eye'. It is caused by various forms of bacterial and viral infections. It includes swimming pool conjunctivitis and the type which is developed as a result of exposure to ultraviolet rays.

Eczema

This is a term now used synonymously with dermatitis. It is not so much a disease as a complex symptom with many causes and clinical variations. There are a number of types or categories, their names —in many cases—indicating their suspected environmental or physiological cause. For example— lipstick dermatitis, perfume dermatitis, housewife's eczema (hand eczema) and industrial dermatitis.

Keloids

These are rather like elevated scars and can occur after burns, cuts, scalds or surgical wounds anywhere on the skin

Psoriasis

Psoriasis is a common form of skin disease estimated to affect some five or six per cent of the population. It affects both sexes and is more commonly found in adults than children. It occurs in families sometimes and in about 25% of cases it is hereditary. Psoriasis is characterised by the development of elevated reddish patches covered by a thick, dry, silvery scale.

Stye or Hordeolum

A stye or hordeolum is usually caused by bacteria getting into the roots of one or more of the eyelashes, where local infection takes place. If, on the other hand, the infection gets into the sweat glands of the eyelid — a *cyst* or *chalazion* forms.

Tineacapitis
(Ringworm of the Scalp)

This superficial fungus infection of the scalp occurs in male or female before.puberty and is characterised by partial loss of scalp hair and the breaking off of infected hairs. It is spread by direct contact with an infected person or through the use of a comb or headgear that has been worn by an infected person..

Verruca (Warts)

There are many types of verruca, the best known being *verruca vulgaris* or common wart and *verruca plantaris* which occurs on the soles of the feet. They are a dry, elevated lesion which may appear singly or in large numbers. They are caused by viruses.

GLOSSARY

Acne Rosacea	unrelated to common acne; a chronic disease which affects the skin of the middle third of the face. It occurs most frequently in middle-aged women in whom it is associated with intestinal disturbances or pelvic disease.
Antrum	a cavity or hollow space in a bone
Auditory Vertigo	dizziness due to disease of the ears
Blister	a collection of fluid between the epidermis and the dermis
Colour Blindness	a congenital inherited condition passed down by the female carriers to their sons
Hirsutism	a condition characterised by growth of hair in unusual places and in unusual amounts
Ocular Vertigo	dizziness due to disease of the eyes
Tinea Pedis (Athlete's Foot)	a fungal infection which is more common in men and more frequent in summer. In the acute form the blister type is the most common.
Vertigo	giddiness; sensation of loss of equilibrium
Vesicle	a small sac containing fluid—a skin blister

Chapter 11

Histology

Histology can be readily defined in two words—microscopic anatomy. It is the branch of biology which deals with the minute structure of tissues which are the basis of cell life.

All living structures are composed of cells and intercellular material. Some of this intercellular material provides strength, for example, collagen and elastic fibres in the skin and calcium salts in the bone, whilst much of the intercellular material acts as a cement between the cells. This is sometimes referred to as *ground substance* or *interstitial substance*.

Life starts when a single ovum (female sex cell) is fertilised by a spermatozoon (male sex cell). These sex cells are formed by a process, meiosis, in which the number of *chromosomes* (genetic material) in the nucleus is halved.

This fertilised cell consists of a nucleus, containing the full complement of chromosomes, surrounded by protoplasm and enclosed by a membrane. It divides by a process called mitosis, in which the essential elements of the nucleus, the chromosomes, are reproduced in each daughter cell. The chromosomes are made up of a linear arrangement of genes. It is now known that the genes in each cell (*genome* contain a complete pattern of the human body.

In 1943 it was discovered that genes were made from very long, large molecules of *deoxyribonucleic acid*, DNA for short, but the way in which DNA carried all the information to produce a complete human being remained unknown until 1953. It was then discovered that DNA was made of four different small molecules called *nucleotides* linked together in a long chain. The most important feature of DNA is that two of these very long chains twist around each other to form a double helix rather like a rubber ladder twisted around its long axis. The DNA molecule is by far the largest molecule found in the cell, which is not surprising, considering the amount of information it has to carry in its four letter alphabet.

In each cell there are something like 30 000 million such letters, equivalent to 1000 books of 1000 pages each. It will be seen that the genes carry the determining factors of inheritance and cell behaviour.

Gradually, as a result of mitosis, a ball of cells is formed and in the very early stages this ball of cells can be divided into three layers:

(1) An outer layer—the *ectoderm* or *epiblast* from which the skin, its appendages and the nervous system are developed.

(2) A middle layer—the *mesoderm* or *mesoblas* from which fat and various internal organs are developed.

3. An inner layer—the *endoblast* which provides the lining of a number of organs of the body.

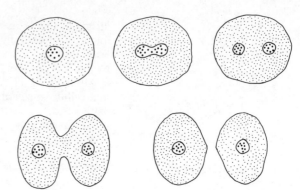

CELL MITOSIS

So the single cell has developed into tissue and, when fully developed, there are four types of tissue in the body:
(1) *The Epithelium*
(2) *Connective Tissue*
(3) *Muscular Tissue*
(4) *Nerve Tissue*

Epithelium

The epithelium is divided into two principal varieties:
a. *Simple epithelium* which consists of one layer of cells. It is very delicate and is found in several organ linings like the thorax and the abdomen.
b. *Stratified* or *compound epithelium* which consists of two or more layers.

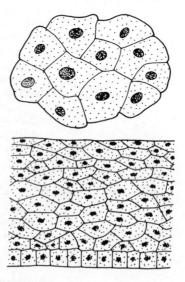

Connective Tissue

This connects all other tissues and when presented in the form of bone gives support and rigidity to the body. There are seven principal varieties of connective tissue:

CONNECTIVE TISSUE

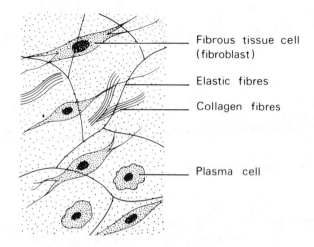

Fibrous tissue cell (fibroblast)

Elastic fibres

Collagen fibres

Plasma cell

(1) Areolar or loose connective tissue—this forms a very thin transparent tissue which surrounds vessels, nerves and muscle fibres.

(2) Adipose tissue—this is not unlike loose connective tissue but the spaces of the network are filled in with fat cells.

FAT CELLS BOUNDED BY WHITE CONNECTIVE TISSUE FIBRES

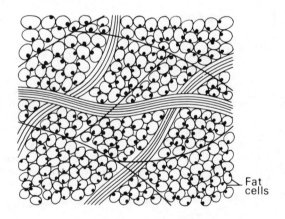

Fat cells

(3) Fibrous tissue—this is found in tendons and ligaments and also forms the outside of various organs such as the kidney and heart, as well as the outside of bone and muscle.

(4) Bone—this is a special type of fibrous material hardened by the deposit of such salts as calcium phosphate, the fibrous material giving it toughness and the mineral matter giving it rigidity.

(5) Cartilage—this is a specialised type of fibrous tissue. It is tough and pliable and very strong. It provides a firm wall to the larynx and the trachea

and for the pads which join bone to bone in the slightly movable joints, for example between the vertebrae.

(6) Yellow elastic tissue—found where elasticity is important as in the walls of blood vessels.

(7) Lymphoid or reticular—found in lymph nodes and the spleen.

WHITE
FIBROCARTILAGE

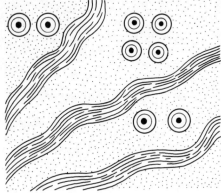

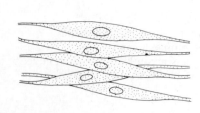

SMOOTH MUSCLE FIBRE

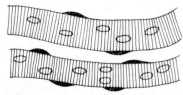

STRIATED MUSCLE FIBRE

Muscular tissue

This is contractile tissue and is able to produce movement. This is dealt with in more detail in the chapter 'The Muscular System'.

Nerve tissue

This has the special function of carrying messages or stimuli throughout the body. It consists of nerve cells and nerve fibres and is dealt with in greater detail under the heading 'The Neurological System'.

NERVE TISSUE

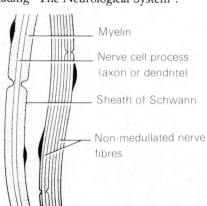

Myelin

Nerve cell process
(axon or dendrite)

Sheath of Schwann

Non-medullated nerve
fibres

In addition to these there are three liquid or fluid tissues—the *blood, lymph* and *cerebro-spinal fluid*. Each of these is dealt with under the appropriate heading.

Membranes

These are made of connective tissue and line the cavities and hollow organs of the body. They secrete lubricating fluids to moisten their smooth surfaces and prevent friction. There are three types of membrane found in the human body:

(1) Synovial membrane secretes a very thick fluid rather like egg white in consistency.

(2) Mucous membrane secretes a sticky fluid called *mucus*.

(3) Serous membrane is made of flattened cells through which a small quantity of a thin, watery substance oozes. The fluid which emanates from it is called *serum*.

Waterlogging of the tissues can occur and when this happens it is called *oedema*. This can arise from a number of causes:

(1) Too high hydrostatic pressure in the capillaries.

(2) An osmotic pressure that is too low.

(3) A blockage of lymphatic vessels.

(4) Damage to the capillary walls.

Cardiac oedema is the type which occurs in congestive heart trouble and this is caused by the increase in venous pressure and the consequent increase in capillary pressure. It is characterised by swelling of the legs and feet of those who habitually walk or stand whilst it appears in the lower part of the back or buttocks of those who lie. An important factor is the kidneys and their diminished secretion of sodium.

SUMMARY

Human beings start as a single cell formed by the fusion of 2 sex cells.

Mitosis is the principle or method used by the cell for multiplication. Most tissues are dealt with under their respective headings but this chapter has primarily been concerned with connective tissue which serves as the supporting system of the body. Its cells are responsible for the elements and matrix of bone, cartilaginous and fibrous tissue. They form the various tough frameworks of the body whilst the matrix supplies lubricating elements which facilitate easy movement.

Collagen was, at one time, used synonymously with connective tissue but is now used more specifically to indicate connective tissue fibres.

Ground substance is a type of cementing material found between the minute fibres and binding them together.

GLOSSARY

Centrosome	a very small, dense part of the cytoplasm lying close to the nucleus
Chromosomes	thread-shaped bodies consisting of DNA, found in the nucleus of the cell. There are 23 pairs (46 in total) in each human cell
Genes	hereditary determinants occurring in the chromosomes in linear arrangement
Grand Cytoplasm	a type of protoplasm surrounding the nucleus, in which the other structures are embedded
Golgi Apparatus	a canal-like structure lying close to the nucleus; named after the famous Italian histologist who first described it at the beginning of this century
Karyokinesis	same as mitosis—cell division
Meiosis	the type of cell division which takes place in the sex organs. The number of chromosomes is halved so that a spermatazoon provides 23 chromosomes and the ovum 23 chromosomes
Mitochondria	small, rodlike structures embedded in the cytoplasm
Pathology	a branch of science which deals with the nature of disease through the study of cause, process, effect and associated alterations of structure and function

GENERAL GLOSSARY

Acute	having rapid onset, a short course with pronounced symptoms
Anastomosis	the intercommunication of the vessels of any system with one another
Ankylosis	stiffness or fixation of a joint
Aplasia	absence of growth
Aponeurosis	the white, shiny membrane covering the muscles or connecting the muscles and tendons with the parts they move
Asthenia	the absence of strength in the muscles
Astigmatism	a defect in the focusing apparatus of the eyes
Benign	a tumour which is not malignant
Bilateral	on both sides
Biopsy	microscopic examination of tissue taken from a living subject
Calorie	a unit of heat that is the amount which raises the temperature of 1 kilogramme of water 1°C
Cardiologist	medically qualified person who specialises in the diagnosis and treatment of disorders of the heart and vascular system
Catalyst	a substance which greatly increases the rate of chemical reaction
Chronic	of long duration—opposite to acute
Condyle	rounded projection at the end of a bone, forming part of a joint with another bone
Costal	relating to the ribs
Digit	finger or toe
Displasia	disordered growth
Dorsum	any part corresponding to the back, e.g. dorsum of the tongue
Dyspnoea	difficulty of breathing
Enzyme	a catalytic substance having a specific action in promoting a chemical change
Effusion	an abnormal outpouring of fluid (serum, pus or blood) into the tissues or cavities of the body)
Electrocardiogram (ECG)	a graphic recording of the electric potential differences due to cardiac action taken from the body surfaces
Electroencephalogram (EEG)	a graphic recording of the minute changes in the electric potential associated with brain activity as detected by electrodes applied to the scalp surface
Filiform	slender—like a thread
Follicle	a small tubular or sac-like depression

Foramen	a hole
Fungiform	having a shape similar to that of a mushroom
Fusiform	spindle-shaped, tapering both ways like a spindle
Gallstones	constituents of gall bladder which have crystallised
Glomerulus	a cluster of capillary vessels in the kidney
Hemiplegia	paralysis of half the body divided vertically
Hepatic	pertaining to the liver
Ingestion	the taking in of food
Keratin	proteins, the chief constituent of nails and hair
Lobe	a rounded part or projection of an organ
Lobule	a small lobe
Matrix	that which encloses anything
Melanin	pigment found in the cells of the skin
Micturition	the act of passing urine
Mitral Valve	the heart valve between the atrium or auricle and ventricle
Myopia	short or near sight
Neoplasm	new growth
Neuroglia	connective tissue of the central nervous system
Neurologist	medically qualified person specialising in the diagnosis and treatment of disorders of the nervous system
Olfactory	pertaining to the sense of smell
Optic	relating to the sense of vision
Orthopaedic Surgeon	a person who is medically qualified and specialises in that branch of surgery devoted to the prevention and correction of bone deformities
Osmosis	the diffusion of liquid substances through membranes
Osseous	bony or composed of bones
Paraplegia	paralysis of half the body divided horizontally
Pathogenic	disease producing
Periphery	circumference; an external surface; the parts away from the centre
Peritoneum	the serous membrane lining the interior of the abdominal cavity
Physiatrist	a person who is specially trained in physical therapy especially for the promotion of health and fitness
Physiotherapist	a person who is specially trained in the science and art of physical medicine
Plexus	an intricate network of nerves, veins or lymphatic vessels

Psychiatrist	a person who is medically qualified and has specialised in mental or psychological conditions
Psychology	the science of the study of the structure and function of the mind; the behaviour of an organism in relation to its environment
Psychosomatic	a body/mind relationship
Pus	the thick white, yellow or greenish fluid found in abscesses, on ulcers or on inflamed and discharging surfaces, composed largely of dead white blood cells
Renal	pertaining to the kidneys
Retina	the innermost and light-sensitive coat of the eyeball
Sarcoma	a malignant tumour composed of cells derived from non-epithelial tissues, mainly connective tissues
Somatic	relating to the body
Squamous	scaly or shaped like a scale
Stenosis	contraction or narrowing of a channel or opening
Symphysis	the line of junction of two bones
Tactile	pertaining to touch or touch sensation
Tricuspid Valve	valve of the heart guarding the passage from the right atrium or auricle to the ventricle
Tuberosity	a protuberance on a bone
Unilateral	on one side
Vasoconstrictor	a nerve causing constriction of a blood vessel
Vasodilator	a nerve causing dilation of a blood vessel

Conclusion

Anatomy and Physiology, e
combined, can provide a fascin
and it is hoped that the forego
some readers to pursue this stud

The acquisition of knowledge i
— when that knowledge encompas
it becomes itself all encompassing.

Index